Student Solutions Manual

Statistics
Learning from Data

Roxy Peck
California Polytechnic State University,
San Luis Obispo

Prepared by

Stephen Miller
Winchester Thurston School, Pittsburgh, PA

BROOKS/COLE
CENGAGE Learning

Australia • Brazil • Japan • Korea • Mexico • Singapore • Spain • United Kingdom • United States

BROOKS/COLE
CENGAGE Learning

ISBN-13: 978-1-285-08983-6
ISBN-10: 1-285-08983-9

Brooks/Cole
20 Channel Center Street
Boston, MA 02210
USA

Cengage Learning is a leading provider of customized learning solutions with office locations around the globe, including Singapore, the United Kingdom, Australia, Mexico, Brazil, and Japan. Locate your local office at:
www.cengage.com/global

Cengage Learning products are represented in Canada by Nelson Education, Ltd.

To learn more about Brooks/Cole, visit
www.cengage.com/brookscole

Purchase any of our products at your local college store or at our preferred online store
www.cengagebrain.com

Printed in the United States of America
1 2 3 4 5 17 16 15 14 13

Contents

Chapter 1 Collecting Data in Reasonable Ways

Section 1.1 Exercise Set 1

1.1: This is an observational study because the person conducting the study merely recorded (based on a survey) whether or not the boomers sleep with their phones within arm's length, and whether or not people ages 50 to 64 used their phones to take photos.

1.2: This is an observational study because the researchers reviewed the history of the children who were participating in the long-term health study. No children were assigned to different experimental groups.

1.3: This is an experiment because the researchers assigned different toddlers to experimental conditions (adult played with/talked to the robot or the adult ignored the robot).

1.4: This is an observational study because the researchers surveyed adult Americans and drew a conclusion from the survey results; there were no experimental treatments assigned.

1.5: This is an experiment because the researchers assigned study participants to one of three treatment groups (meditation, distraction task, or relaxation technique).

Additional Exercises for Section 1.1

1.11: This is an observational study because the researchers observed the proportion of patients who got an infection in the two groups (overnight hospitalization or more than one night hospitalization); there was no assignment of subjects to experimental groups.

1.13: This is an experiment because the study participants were assigned to one of the two experimental groups (how much would you pay for the mug or how much would you sell the mug for).

Section 1.2 Exercise Set 1

1.15: (a) The group of 716 bicycle fatalities represents a census of the 2008 bicycle fatalities. (b) Because the group of 716 represents a census, the average age of 41 years is a population characteristic.

1.16: The sample is the 2,121 children between the ages of 1 and 4, and the population of interest is all children between the ages of 1 and 4.

1.17: No, it is not safe to generalize this result to the larger population of U.S. adults. The 6000 people who sent hair samples were not chosen using a random selection process. Rather, they voluntarily sent their hair samples.

1.18: There are several reasonable approaches. One is described here. Using the list of all students at the school, write their names on identical but different slips of paper. Thoroughly mix the slips of paper, and select 150 slips. The individuals whose names are on the slips of paper constitute the sample.

1.19: (a) The population of interest is all U.S. women. (b) Although the details of the sampling scheme are not presented, the sample size is large (which is generally desirable). However, not all states were represented in the sample; only women from Maryland, Minnesota, Oregon and Pennsylvania were included in the sample. As such, it might be difficult to generalize beyond the population of women in those four states. (c) Given that only women from four states were included in the sample, the sample is not likely to be representative of the population of interest. (d) Selection bias is present because the selection method excluded women from all states other than Maryland, Minnesota, Oregon and Pennsylvania.

Additional Exercises for Section 1.2

1.25: The population is all 7000 property owners in this particular rural county. The sample is the 500 property owners selected at random from the 7000 total owners in the county.

1.27: The population is the 5000 bricks in the lot available at the auction. The sample is the 100 bricks chosen for inspection.

1.29: Bias introduced through the two different sampling methods may have contributed to the different results. The online sample could suffer from voluntary response bias in that perhaps only those who feel very strongly would take the time to go to the website and register their vote. In addition, younger people might be more technologically savvy, and therefore the website might represent the views of younger people (particularly students) who support the parade. The telephone survey telephone responses might represent the view of permanent residents (as students might only use cell phones and not have a local phone number).

Section 1.3 Exercise Set 1

1.30: Random assignment allows the researcher to create groups that are equivalent, so that the subjects in each experimental group are as much alike as possible. This ensures that the experiment does not favor one experimental condition (playing Unreal Tournament 2004 or Tetris) over another.

1.31: (a) Allowing subject participants to choose which group they want to be in could introduce systematic differences between the two experimental conditions (tai chi group or control group), resulting in potential confounding. Those who would choose to do tai chi might, in some way, be different from those who would choose the control group. We would not know if differences in immunity between the two groups were due to the tai chi, or due to some inherent differences in the subjects who chose their experimental groups. (b) Because the purpose of this experiment is to determine whether the tai chi treatment has an effect on immunity to a virus, a control group is needed to provide a baseline against which the treatment group can be compared to determine if the treatment has an effect.

1.32: (a) The attending nurse was responsible for administering medication after judging the degree of pain and nausea, so the researchers did not want the nurse's personal beliefs about the different surgical procedures to influence measurements. (b) Because the children who had the surgery could easily determine whether the surgical procedure was laparoscopic repair or open repair based on the type of incision.

1.33: There are several possible approaches. One is described here. Write each subject's name on identical slips of paper. Mix the slips of paper thoroughly and draw out slips one at a time. The names on the first 15 slips are assigned to the experimental condition of listening to a Mozart piano sonata for 24 minutes. The names on the next 15 slips are assigned to the experimental condition of listening to popular music for the same length of time. The remaining 15 names are assigned to the relaxation with no music experimental condition.

1.34: (1) Do ethnic group and gender influence the type of care that a heart patient receives? (2) The experimental conditions are the gender and race of the "patient" the doctor is shown. (3) The response variable is the type of care recommended for the heart patient. (4) The experimental units are the 720 primary care doctors at meetings of the American College of Physicians or the American Academy of Family Physicians. It is not clear how the physicians were chosen. (5) Yes, the design incorporates random assignment of doctors to view one of the four different videos through rolling a four-sided die. (6) No, there was no control group, as all the doctors were shown actions of some race or gender. The idea of a control group does not apply in this study. (7) There is no indication that the study includes blinding. There cannot be blinding in this study because the doctors know the gender and race of the "patients" they get.

Additional Exercises for Section 1.3

1.41: (a) Some surgical procedures are more complex and require a greater degree of concentration; music with a vocal component might be more distracting when the surgical procedure is more complex. (b) The temperature of the room might affect the comfort of the surgeon; if the surgeon is too hot or too cold, she or he might be uncomfortable, and therefore more easily distracted by the vocal component. (c) If the music is too loud, the surgeon might be distracted and unable to focus, regardless of the presence or absence of the vocal component. If the music is too soft, the surgeon might try to concentrate on listening to the vocal component, and therefore pay more attention to the music rather than the surgical procedure.

1.43: This experiment could not have been double-blind because the surgeon would know whether or not there was a vocal component to the music.

1.45: (a) Probably not, because the judges might not believe that Denny's food is as good as other restaurants. (b) Experiments are often blinded in this way to eliminate preconceptions about particular experimental treatments.

Section 1.4 Exercise Set 1

1.46: It is not reasonable to conclude that being raised with two or more animals is the cause of the observed lower allergy rate. This was an observational study, so cause-and-effect conclusions cannot be drawn.

1.47: (a) It is not reasonable to conclude that watching *Oprah* causes a decrease in cravings for fattening foods. This was an observational study, so cause-and-effect conclusions cannot be drawn. (b) It is not reasonable to generalize the results of this survey to all women in the United States because not all women watch daytime talk shows. It is not reasonable to generalize these results to all women who watch daytime talk shows because not all women who watch daytime shows access DietSmart.com. If there was no random selection of survey participants (which is often the case with surveys found on websites), then the results might be biased due to voluntary response of participants.

1.48: The researcher would have had to assign the nine cyclists at random to one of the three experimental conditions (chocolate milk, Gatorade, or Endurox).

1.49:

Study 1:

Question 1: This is an observational study.

Question 2: Yes, random selection was used.

Question 3: No, this was not an experiment so there were no experimental groups.

Question 4: No, because this was not an experiment, cause-and-effect cannot be concluded.

Question 5: It is reasonable to generalize to the population of students at this particular large college.

Study 2:

Question 1: This study was an experiment.

Question 2: Random selection was not used.

Question 3: There was no random assignment to experimental conditions (the grouping was based on gender).

Question 4: No, the conclusion is not appropriate because of confounding of gender and treatment (women ate pecans, and men did not eat pecans).

Question 5: It is not reasonable to generalize to a larger population.

Study 3:

Question 1: This is an observational study.

Question 2: There was no random selection.

Question 3: There was no random assignment to experimental groups.

Question 4: No, the conclusion is not appropriate because this was an observational study, and therefore cause-and-effect conclusions cannot be drawn.

Question 5: We cannot generalize to any larger population beyond the 200 volunteers.

Study 4:

Question 1: This is an experiment.

Question 2: There was no random selection from some population.

Question 3: Yes, there was random assignment to experimental groups.

Question 4: Yes, because this was a simple comparative experiment with random assignment of subjects to experimental groups. We can draw cause-and-effect conclusions.

Question 5: We cannot generalize to a larger population because there was no random selection from some population.

Study 5:

Question 1: This is an experiment.

Question 2: Yes, there was random selection from students enrolled at a large college.

Question 3: Yes, random assignment of subjects to experimental groups was used.

Question 4: Yes, because this was a simple comparative experiment with random assignment of subjects to experimental groups. We can draw cause-and-effect conclusions.

Question 5: Due to the random selection of students, we can generalize conclusions from this study to the population of all students enrolled at the large college.

Additional Exercises for Section 1.4

1.55: In order to determine if the conclusions implied by the headline are appropriate, I would need to know if dieters were randomly assigned to the experimental conditions (large fork or small fork). In order to generalize to the population of dieters, I would also want to know if the study participants were randomly selected from the population of dieters.

1.57: There was no random selection from some population.

1.59: Yes, it is reasonable to draw the conclusion that reasoning using information displayed on a small screen is improved by turning the screen to landscape orientation because this was an experiment in which there was random assignment of subjects to experimental groups.

Are You Ready to Move On? Chapter 1 Review Exercises

1.61: (a) This is an experiment due to the random assignment of subjects to experimental conditions (the five different rooms). (b) This is an observational study because there was no random assignment of subjects to experimental conditions; the researchers merely recorded what they observed on the MySpace pages. (c) This is an observational study because there was no random assignment of subjects to experimental conditions; the researchers merely recorded the responses of the survey participants. (d) This is an experiment because of the random assignment of study participants (the adults with back pain) to experimental conditions (the four different treatments).

1.63: (a) 84% is a population characteristic. (b) 24.1 years is a statistic. (c) 22% is a population characteristic. (d) 6.4 days is a statistic. (e) 63 hours is a statistic.

1.65: The council president should assign a unique identifying number to each of the names on the petition, numbered from 1 to 500. On identical slips of paper, write the numbers 1 to 500, with each number on a single slip of paper. Thoroughly mix the slips of paper and select 30 numbers. The 30 numbers correspond to the unique numbers assigned to names on the petition. These 30 names constitute the sample.

1.67: Without random assignment of the study participants to experimental condition, confounding could impact the conclusions of the study. For example, people who would choose an attractive avatar might be more outgoing and willing to engage than someone who would choose an unattractive avatar.

1.69: (a) It seems as if the alternate assignment to the experimental groups (large serving bowls, small serving bowls) would tend to produce groups that are similar. People who arrive to the party at approximately the same time might, in some way, be similar to each other, so dividing them into the different experimental groups as described would tend to make the two groups similar to each other. (b) Blinding ensures that individuals do not let personal beliefs influence their measurements. The research assistant that weighed the plates and estimated the calorie content of the food might (intentionally or not) have let her or his personal beliefs influence the estimate of the calorie content of the food on the plate.

1.71: (a) (1) The experiment is designed to answer the question "Does using hand gestures help children learn math?" (2) The two experimental conditions are using hand gestures and not using hand gestures. (3) The response variable is the number correct on the six-problem test. (4) The experimental units are the 128 children in the study; they were selected because they were the children who answered all six questions on the pretest incorrectly. (5) Yes, the children were assigned randomly to one of the two experimental groups. (6) Yes, the control group is the experimental condition of not using any hand gestures. (7) There was no blinding and, indeed, it would not be possible to include blinding of subjects in this experiment (the children would know whether or not they were using hand gestures), and there is no need to blind the person recording the response because the test was graded with each answer correct or incorrect, so there is no subjectivity in recording the responses. (b) It seems as if the conclusions are reasonable because the subjects were assigned to the treatment groups at random.

1.73: (a) No, the 60 games selected were the 20 most popular (by sales) for each of three different gaming systems. The study excluded the games that were not in the top 20 most popular (by sales). (b) It is not reasonable to generalize the researcher's conclusions to all video games due to selection bias (there was a systematic exclusion of those games not in the top 20 most popular (by sales)).

1.75:

Study 1:

Question 1: The study described is an observational study.

Question 2: No, there was no random selection from a population.

Question 3: No, there was no random assignment to experimental groups.

Question 4: No, it is not reasonable to conclude that taking calcium supplements is the cause of the increased heart attack risk.

Question 5: No, it is not reasonable to generalize conclusions from this study to a larger population because there was no random selection from a larger population.

Study 2:

Question 1: The study described is an observational study.

Question 2: Yes, there was random selection from the population of people living in Minneapolis who receive Social Security.

Question 3: No, there was no random assignment of subjects to experimental groups.

Question 4: No, it is not reasonable to conclude that taking calcium supplements is the cause of the increased heart attack risk.

Question 5: Yes, it is reasonable to generalize the results of this study to the population of people living in Minneapolis who receive Social Security.

Study 3:

Question 1: The study described is an experiment.

Question 2: Yes, there was random selection from the population of people living in Minneapolis who receive Social Security.

Question 3: No, there was no random assignment of subjects to experimental groups.

Question 4: No, it is not reasonable to conclude that taking calcium supplements is the cause of the increased risk of heart attack due to confounding and the lack of random assignment of subjects to experimental conditions. The participants in this study who did not have a previous history of heart problems were given the calcium supplement, and those with a history of heart problems were not given the supplement. It is not possible to determine the role of the calcium supplement because only those study participants who did not have a history of heart problems were given the supplement.

Question 5: It is possible to generalize the results from this study to the population of all people living in Minneapolis who receive Social Security. However, it is unclear (due to the confounding described in Question (4) what the conclusion would be.

Study 4:

Question 1: The study described is an experiment because there was random assignment of subjects to experimental conditions.

Question 2: No, there was no random selection from some larger population.

Question 3: Yes, there was random assignment of study participants to experimental groups.

Question 4: Yes, it is reasonable to conclude that taking calcium supplements is the cause of the increased risk of heart attack.

Question 5: No, it is not reasonable to generalize conclusions from this study to some larger population because of the lack of random selection of study participants from a population.

Chapter 2 Graphical Methods for Describing Data Distributions

Section 2.1 Exercise Set 1

2.1: (a) numerical, discrete (b) categorical (c) numerical, continuous (d) numerical, continuous (e) categorical

2.2: (a) discrete (b) continuous (c) discrete (d) discrete

2.3:

Data Set 1:

Question 1: There is one variable in the data set.

Question 2: The data set is categorical.

Question 3: The purpose of the graphical display is to summarize the data distribution.

Appropriate Graphical Display: Bar chart

Data Set 2:

Question 1: There is one variable in the data set.

Question 2: The data set is numerical.

Question 3: The purpose of the graphical display is to compare groups (full-time students or part-time students).

Appropriate Graphical Display: Comparative dotplot, comparative stem-and-leaf displays, and comparative histograms are all appropriate.

Data Set 3:

Question 1: There are two variables in the data set.

Question 2: The data set is numerical.

Question 3: The purpose of the graphical display is to investigate the relationship between two numerical variables.

Appropriate Graphical Display: Scatterplot

Data Set 4:

Question 1: There is one variable in the data set.

Question 2: The data set is categorical.

Question 3: The purpose of a graphical display is to compare groups (faculty, students)

Appropriate Graphical Display: Comparative bar chart

Data Set 5:

Question 1: There is one variable in the data set.

Question 2: The data set is numerical.

Question 3: The purpose of a graphical display is to summarize a data distribution.

Appropriate Graphical Display: Dotplot, stem-and-leaf, and histogram are all appropriate graphical displays.

Additional Exercises for Section 2.1

2.7: (a) numerical (b) numerical (c) categorical (d) numerical (e) categorical

2.9: (a) categorical (b) numerical (c) numerical (d) categorical

2.11: (a) numerical (b) numerical (c) numerical (d) categorical (e) categorical (f) numerical (g) categorical

2.13:

Question 1: There is one variable in the data set.

Question 2: The data set is numerical.

Question 3: The purpose is to compare groups (male students, female students).

Appropriate graphical display: Comparative dotplot, comparative stem-and-leaf, or histograms are all appropriate.

2.15:

Question 1: There is one variable in the data set.

Question 2: The data set is numerical.

Question 3: The purpose of a graphical display is to summarize the data distribution.

Appropriate graphical display: Dotplots, stem-and-leaf plots, and histograms are all appropriate graphical displays.

Section 2.2 Exercise Set 1

2.16: (a)

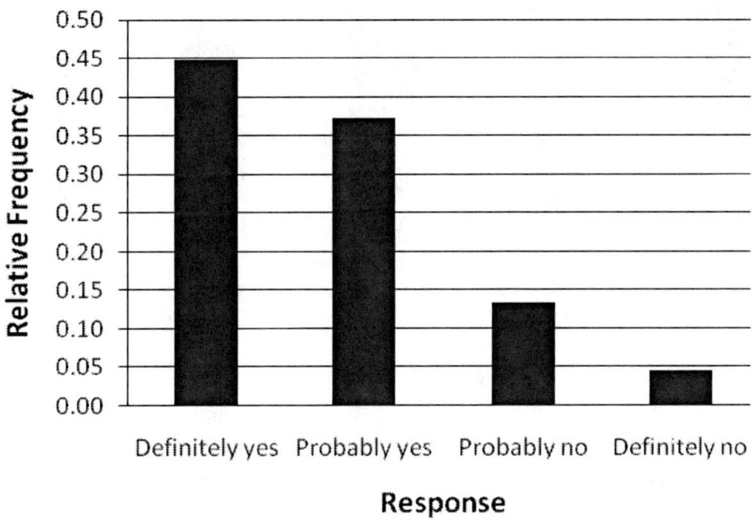

(b) "Senior Satisfaction! Over 80% say they would enroll again."

2.17:

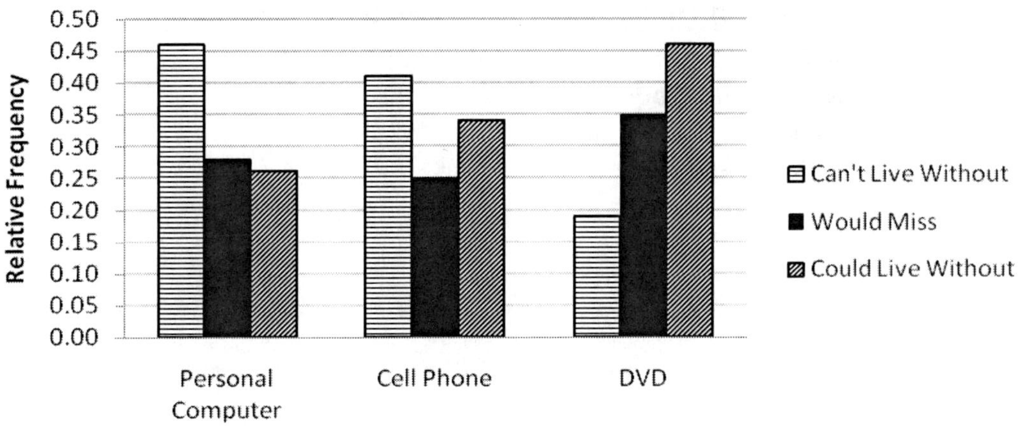

Additional Exercises for Section 2.2

2.21:

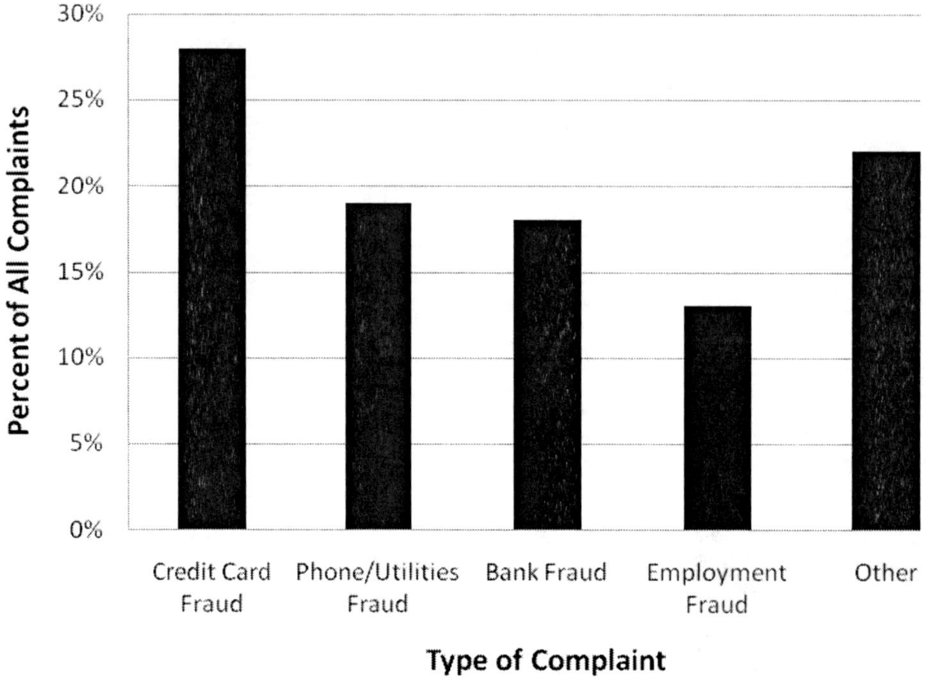

Credit card fraud is the most commonly occurring identity theft type. Although phone/utility, bank, and employment fraud each constitute a relatively large portion of overall type of identity theft, the collective "other fraud" category is greater than any one of these other three.

2.23: The relative frequency distribution is:

Type of Household	Relative Frequency
Nonfamilies	0.29
Married with Children	0.27
Married without Children	0.29
Single Parent	0.15

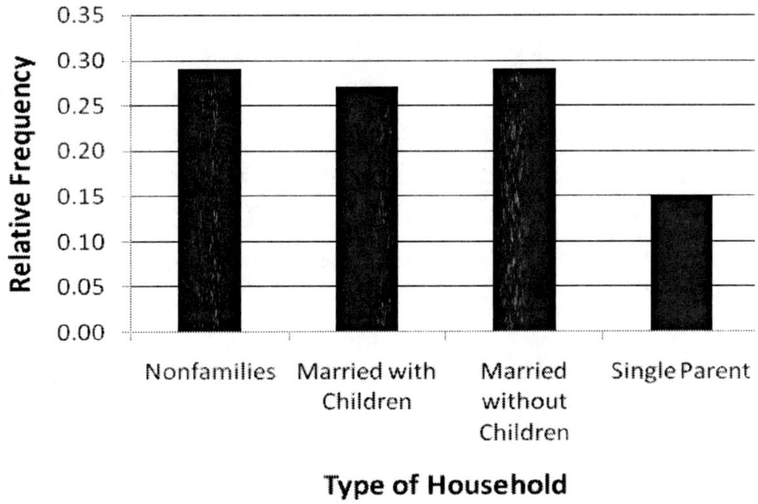

Section 2.3 Exercise Set 1

2.24: (a) The dotplot below shows the cost per gram of protein for 19 common food sources of protein.

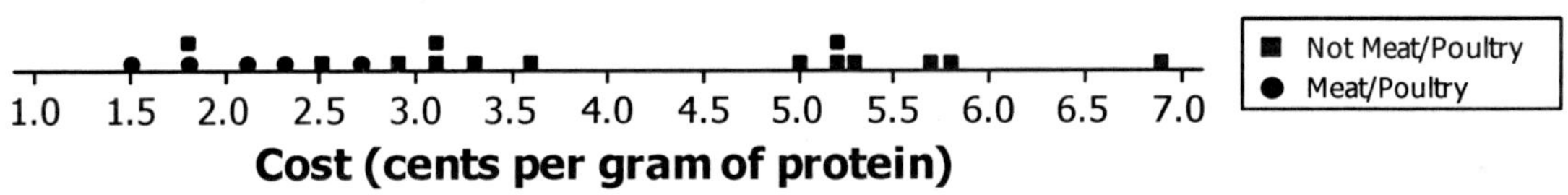

(b) Because the meat and poultry products (represented by the circles on the dotplot) are located at generally smaller values of cost (in cents per gram of protein), they do appear to be a good value when compared to other sources of protein.

2.25: (a)

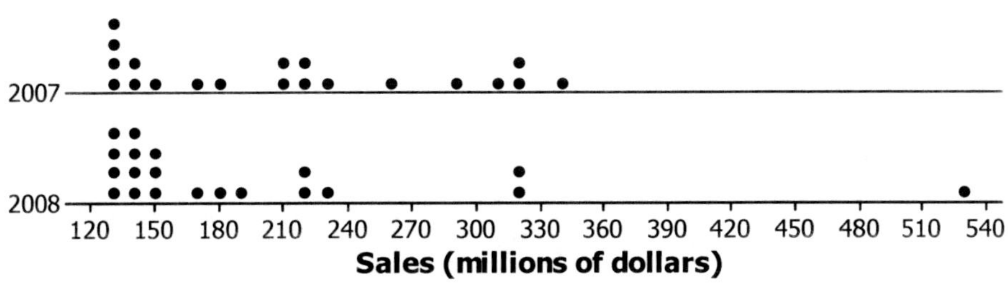

(b) The shape of both distributions is skewed toward larger values. The 2007 ticket sales is centered at about $210 million dollars, which is higher than the center of the 2008 ticket sales, which is centered around $150 million dollars. The lowest ticket sales for both 2007 and 2008 are approximately $127 million dollars. Ticket sales for 2008 has a maximum value of approximately $533 million dollars, which is much higher than the highest ticket sales for 2007. Without this extreme value, the spread between the lowest and highest values are approximately equal, with 2007 having a slightly higher spread than 2008.

2.26:

```
28 | 8
29 |
30 |
31 |
32 | 8
33 | 0
34 | 178
35 | 00145678899
36 | 238999
37 | 0034566777
38 | 01124558
39 | 00259
40 | 045
41 | 2
42 | 2

     Legend: 34|1 = 34.1 years
```

The distribution of median ages is centered at approximately 37 years old, with values ranging from 28.8 to 42.2 years. The distribution is approximately symmetric, with one possible outlier of 28.8 years.

2.27: (a)

Very Large Urban Area		Large Urban Area
	1	023478
	2	369
8	3	0033589
99	4	0366
8711	5	012355
9730	6	
2	7	
	8	
3	9	

Legend: 4|6 = 46 extra hours per year

(b) The statement "The larger the urban areas, the greater the extra travel time during peak period travel" is generally consistent with the data. Although there is overlap between the times for the very large and large urban areas, the back-to-back stem-and-leaf plot shows that for the very large urban areas, the extra travel time during peak period is generally longer than for the large urban areas. The extra travel time during peak period is centered at approximately 58 hours for the very large urban areas, which is higher than the center of approximately 34 hours for the large urban areas.

2.28: (a)

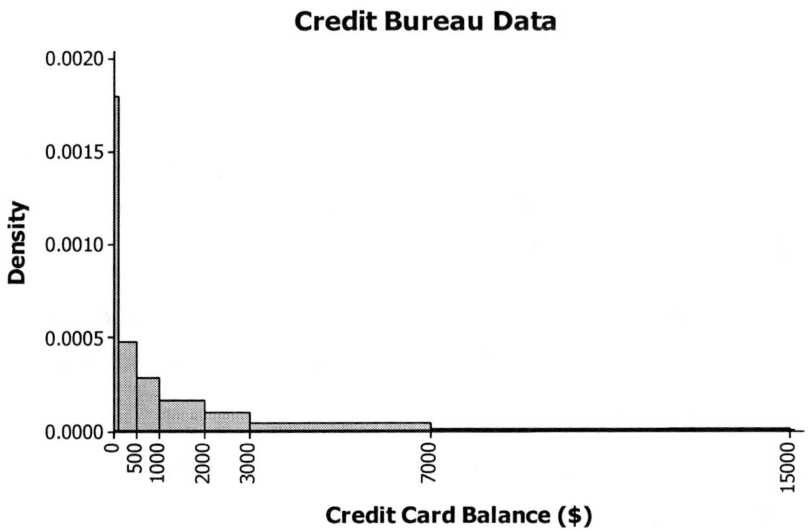

(b)

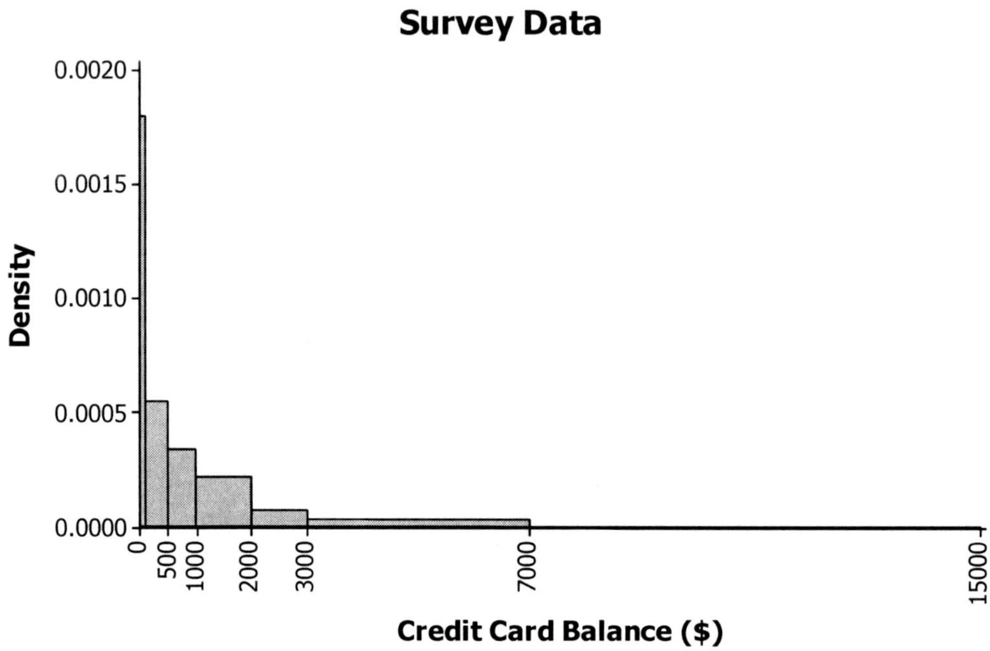

(c) The histograms are similar in shape. A notable difference is that the Credit Bureau data shows that 7% of students have credit card balances of at least \$7000, but no survey respondent indicated a balance of at least \$7000.

(d) Yes, it is likely that the high nonresponse rate for the survey may have contributed to the observed difference in the two histograms because students with credit card balances of \$7000 or more might be too embarrassed to admit that they have such a high balance.

2.29: (a) If the exam is quite easy, the scores would be clustered at the high end of the scale, with a few low scores for the students who did not study. The histogram would be skewed toward low values.

(b) If the exam is difficult, the scores would be clustered around a much lower median value. There might be a few high scores, therefore the histogram would be skewed toward higher values.

(c) In this case, the histogram would be bimodal, with a cluster of high scores and a cluster of low scores.

Additional Exercises for Section 2.3

2.37: The distribution of wind speed is positively skewed and bimodal. There are peaks in the 35-40 m/s and 60-65 m/s intervals.

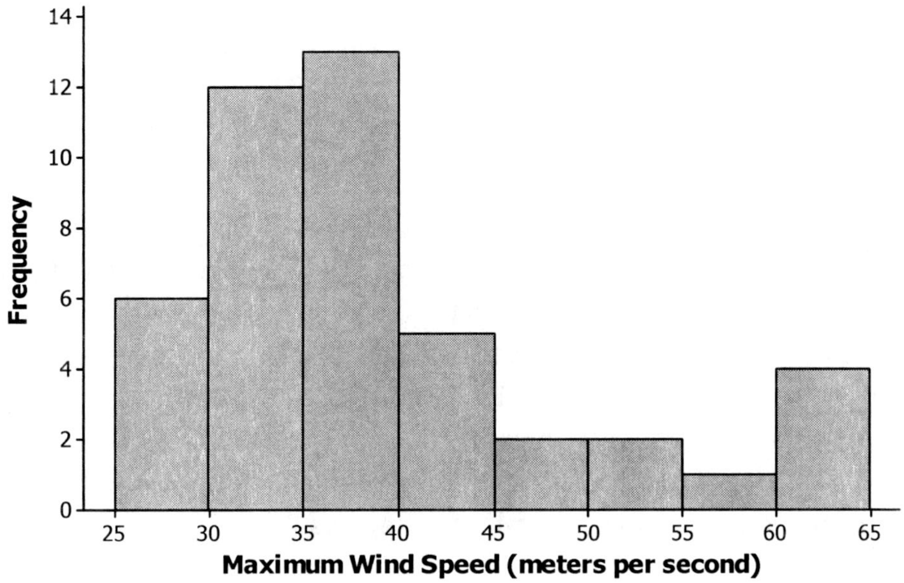

2.39:

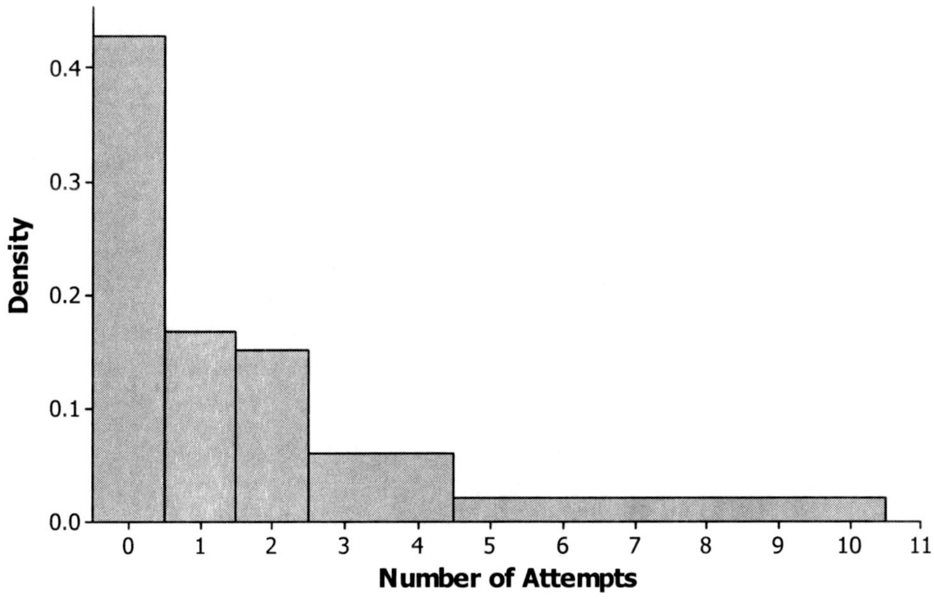

2.41:

Non-Disney		Disney
9521000	0	001233579
651	1	567
0	2	029
	3	
	4	
	5	4

Legend: 1|5 = 150 seconds

In general, the Disney movies have longer tobacco exposure times than the non-Disney movies. Disney movies have a typical value of approximately 80 seconds, which is larger than a typical value for the non-Disney movies of 50 seconds. In addition, there is more variability in the Disney tobacco exposure times than the others as well. Disney movies vary between 6 and 548 seconds, which is greater than the observed spread for the non-Disney movies, which vary between 1 and 205 seconds. Finally, there appears to be one outlier in the Disney movies (548 seconds), and no outliers in the non-Disney movies.

Section 2.4 Exercise Set 1

2.42: (a)

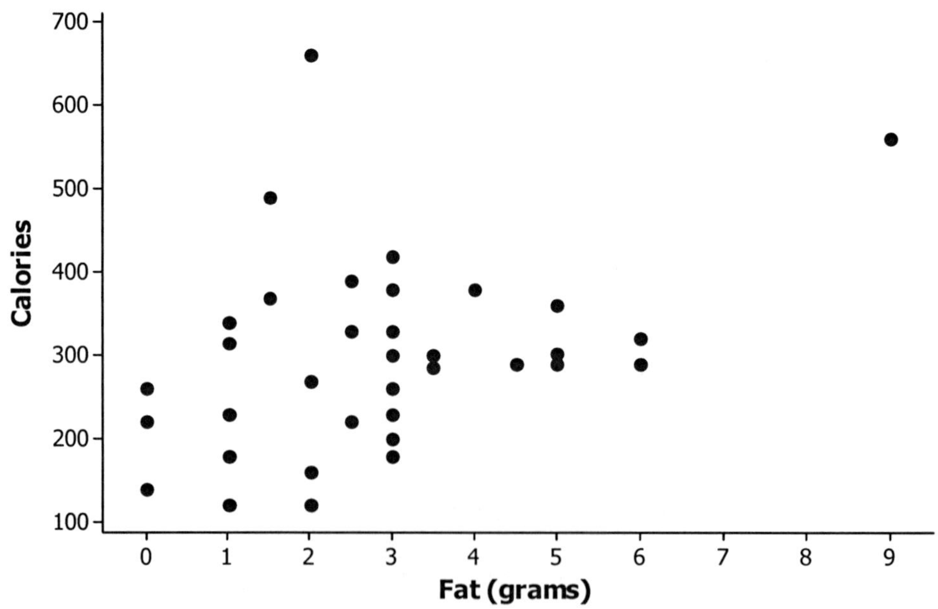

The scatterplot shows the expected positive relationship between grams of fat and calories. The relationship is weak.

(b)

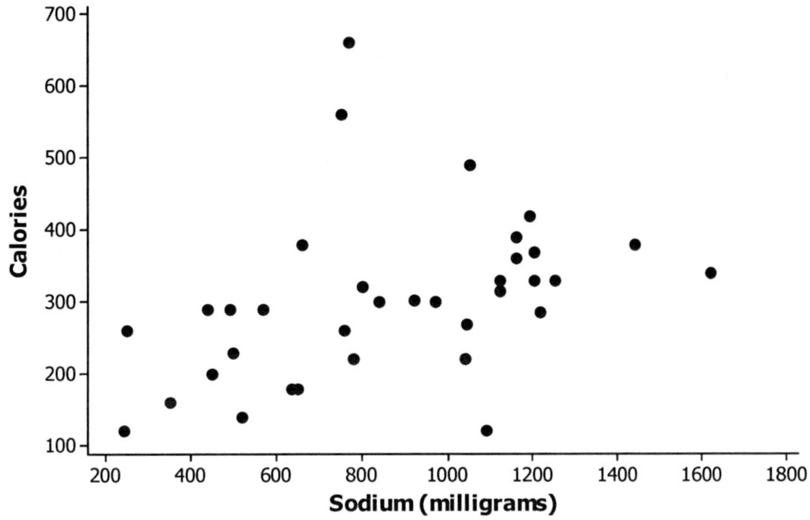

As was observed in the calories vs. fat scatterplot, there is also a weak, positive relationship between calories and sodium. The relationship between calories and sodium appears to be a little stronger than the calories vs. fat relationship.

(c)

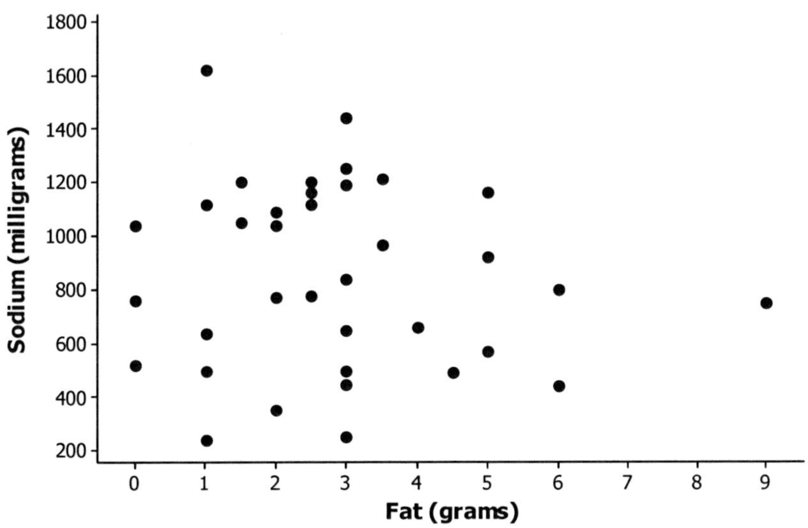

There is no apparent relationship between sodium and fat.

(d)

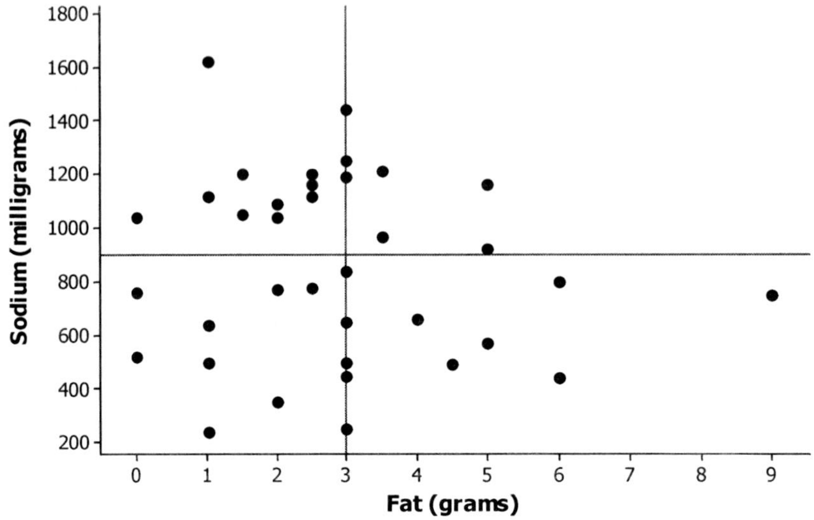

The lower left quadrant corresponds to healthier fast-food choices. This quadrant corresponds to food items with fewer than 3 grams of fat, and fewer than 900 milligrams of sodium, which are considered the healthier choices.

2.43:

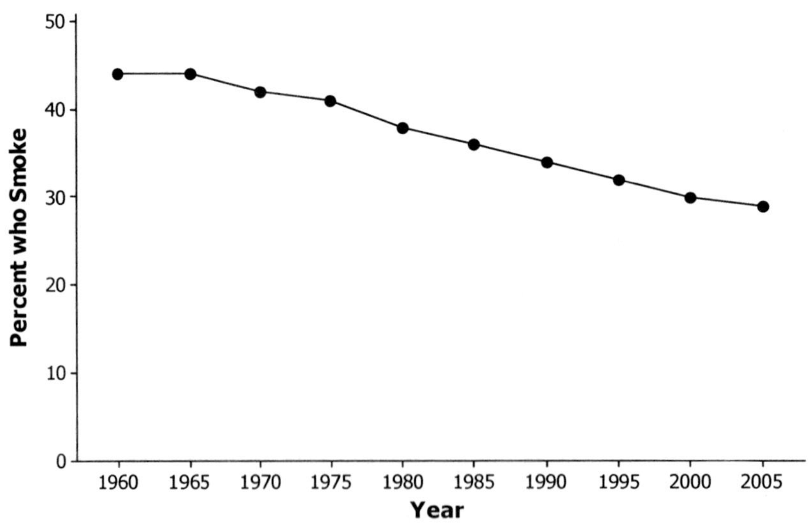

There has been a steady downward trend in the percent of people who smoke among people who did not graduate from high school, from a high of 44% to 29% in 2005.

(b)

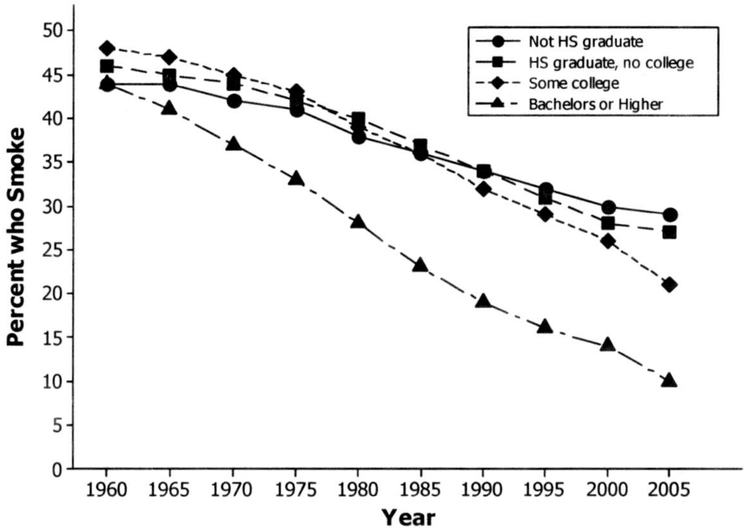

(c) There has been a steady downward trend in the percent of people age 25 or older who smoke regardless of education level. In 1960, regardless of education level, approximately the same percentage of people age 25 or older smoked (approximately 44-48%). Over time, however, the differences between the percent of people who smoke for different education levels has become more pronounced. In 2005, those people with bachelor's degrees or higher had the lowest smoking rate (10%), followed by those with some college (21%). The highest rates of smoking were found among those who either did not graduate from high school (29%), or graduated high school but did not attend college (27%).

Additional Exercises for Section 2.4

2.47: (a)

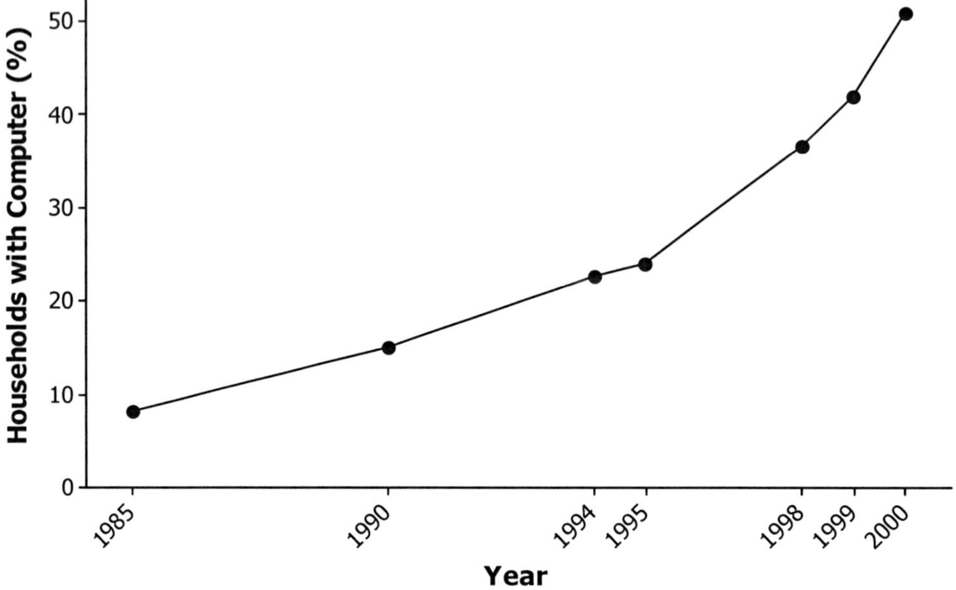

(b) The percentage of households with computers has increased over time, from a low of approximately 8% in 1985 to over 50% in 2000. The rate of increase of the percentage of households with a computer has also increased over time.

2.49: There is a relatively weak positive relationship between poverty rate and dropout rate. There are two states that have poverty rates between 10% and 15%, but have much higher dropout rates (over 15%) compared to other states with comparable poverty rates.

Section 2.5 Exercise Set 1

2.51: (a) The response is a categorical variable.

(b) A bar chart was used rather than a dotplot because the response is a categorical variable, and dotplots are used for numerical responses.

(c) This is not a correct representation of the response data because the percent values add up to over 100% (they add to 107%).

2.52: (a) Overall score is numerical. Grade is categorical. (b) The figure is equivalent to a segmented bar graph because the bar is divided into segments, with different shaded regions representing the different grades ("Top of the Class," "Passing," "Barely Passing," and "Failing"), and the height of each segment is equal to the frequency for that category (for example, there are 5 school districts in the Top of the Class category, 3 in the Passing category, and so on), thus making the area of each shaded region proportional to the relative frequencies for each grade. These conditions satisfy the definition of a segmented bar chart given in the textbook.

(c)

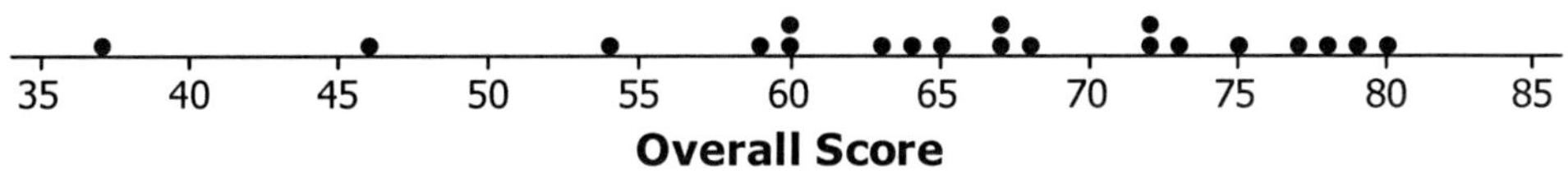

One alternate assignment of grades is to require that "Top of the Class" schools earn grades of 72 or higher, "Passing" schools earn between 66 and 71, "Barely Passing" schools earn between 61 and 65, and "Failing" schools earn 60 or below. This alternative is suggested because there appear to be clusters of dots on the dotplot that correspond to the alternate suggested ranges.

2.53: Answers may vary. For teens age 12 to 17 years, the percentage of cell phone owners increases with age in each of the years 2004, 2006, and 2008. In addition, within each age group, the percentage of teens owning cell phones increased between 2004 and 2008. The largest increase in percentage of teens owning cell phones was among 12 year olds (increasing from 18% to 51% between 2004 and 2008), and the least percentage increase was among 13 year olds (increasing from 34% to 53% between 2004 and 2008).

Additional Exercises for Section 2.5

2.57: (a) The areas in the display are not proportional to the values they represent. The "no" category seems to represent more than 68%.

(b)

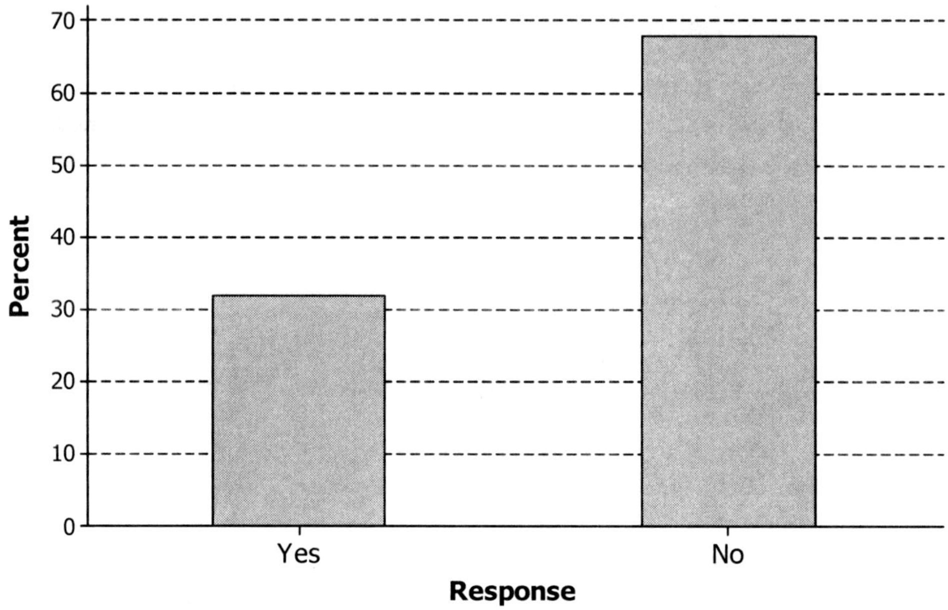
70
60
50
40
30
20
10
0
Percent
Yes
No
Response

2.59: (a)

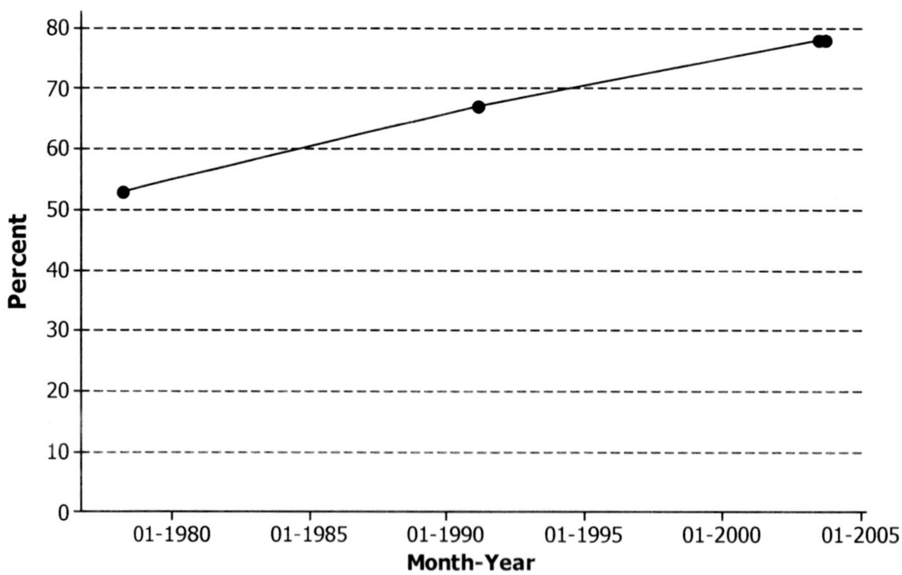
80
70
60
50
40
30
20
10
0
Percent
01-1980
01-1985
01-1990
01-1995
01-2000
01-2005
Month-Year

(b)

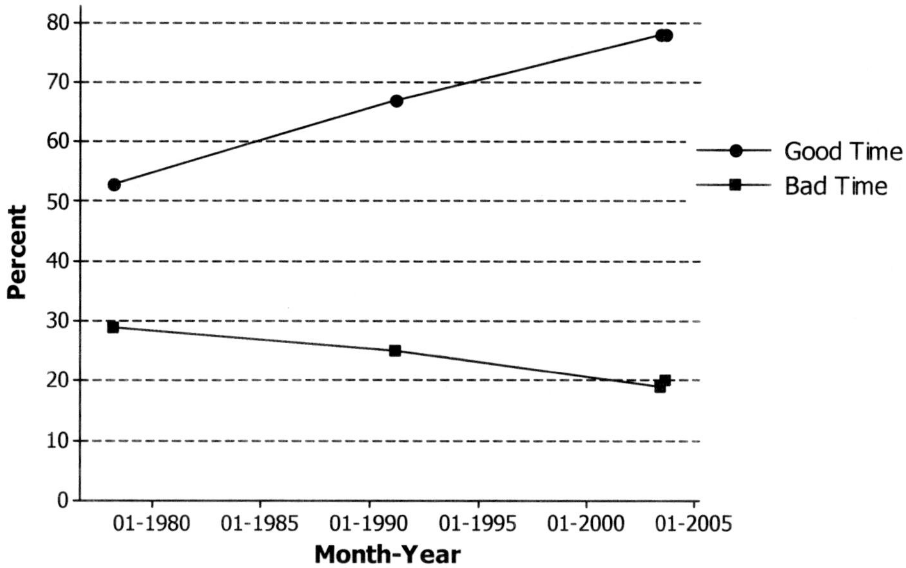

(c) The time-series plot best shows the trend over time.

Are You Ready to Move On? Chapter 2 Review Exercises

2.61:

Data Set 1

Question 1: There is one variable in the data set.

Question 2: The data set is numerical.

Question 3: The purpose of the graphical display is to summarize the data distribution.

Appropriate Graphical Display: Dotplot, stem-and-leaf display, or histogram

Data Set 2

Question 1: There are two variables in the data set.

Question 2: The data set is numerical.

Question 3: The purpose of the graphical display is to investigate the relationship between two numerical variables.

Appropriate Graphical Display: Scatterplot

Data Set 3

Question 1: There are two variables in the data set.

Question 2: The data set is categorical.

Question 3: The purpose of the graphical display is to compare groups.

Appropriate Graphical Display: Comparative Bar Chart

Data Set 4

Question 1: There is one variable in the data set.

Question 2: The data set is categorical.

Question 3: The purpose of the graphical display is to summarize the data distribution.

Appropriate Graphical Display: Bar chart

Data Set 5

Question 1: There is one variable in the data set.

Question 2: The data set is numerical.

Question 3: The purpose of the graphical display is to compare groups.

Appropriate Graphical Display: Comparative dotplot, comparative stem-and-leaf display.

2.63: (a)

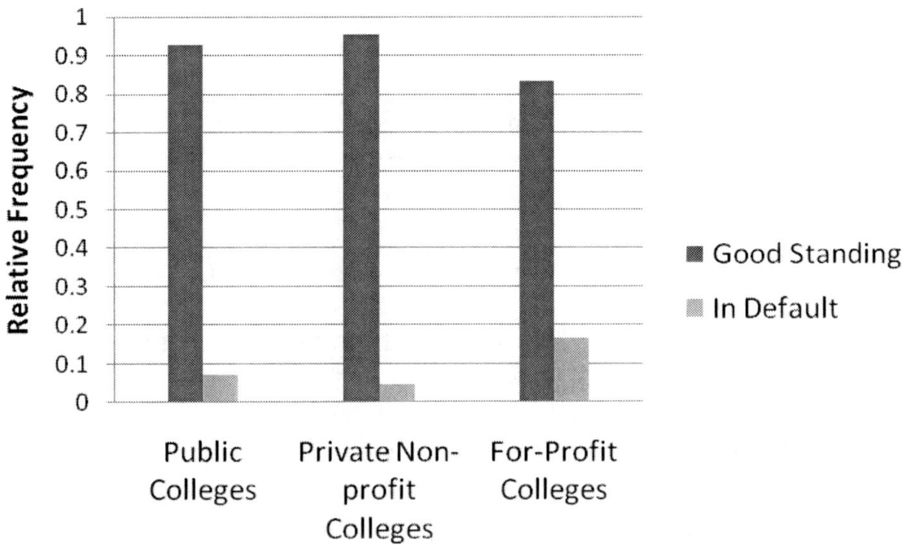

(b) The aspect of the comparative bar chart that supports the statement "those who attended for-profit schools were more likely to default than those who attended public or private non-profit schools" is that the "In Default" bar in the "For-Profit Colleges" category is taller than either of the other "In Default" bars.

2.65: (a)

```
0 | 8
1 | 4467788999
2 | 00001122233333444555777 8
3 | 01222334457
4 | 02458

   Legend: 1|4 = 14 cents per gallon
```

(b) The center is approximately 24 cents per gallon, and most states have a tax that is near the center value, with tax values ranging from 8 cents per gallon to 48 cents per gallon. The distribution is also approximately symmetric.

(c) The only value that might be considered unusual is the 8 cents per gallon tax in Alaska. Although not an apparent outlier, it is nearly 6 cents per gallon lower than the next lowest gasoline tax. There are no other states that make such a big jump to the next higher tax.

2.67: (a)

(b) A typical value for quality rating is approximately 114.

(c) There is a lot of variability, with quality rating ranging between a low score of 84 defects per 100 vehicles and a high score of 170 defects.

(d) Two brands (Land Rover and Mitsubishi) seem to stand out as having a much greater value for number of defects. Four brands (Acura, Lexus, Mercedes-Benz, and Porsche) have smaller values for number of defects.

(e)

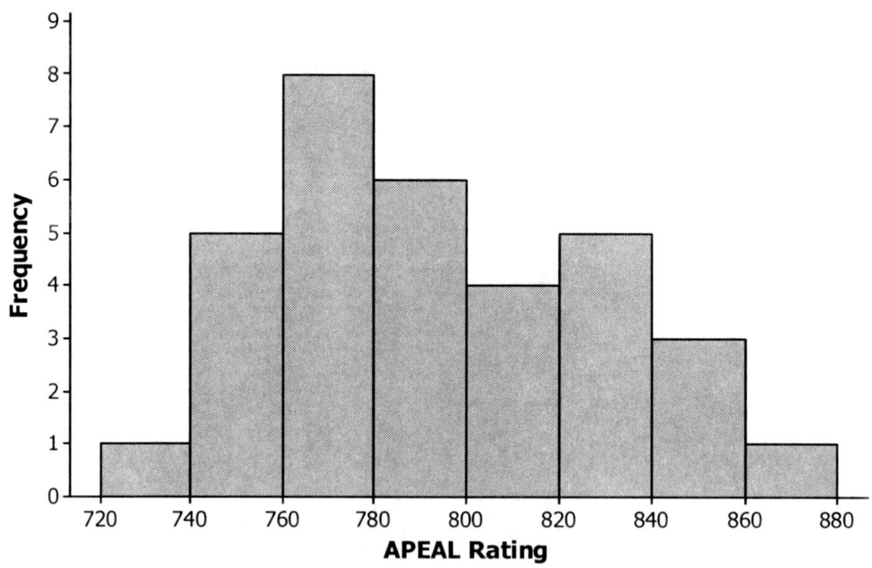

The histogram is centered at approximately 790, with values that range between approximately 720 and 880. The distribution is bimodal and positively skewed.

(f)

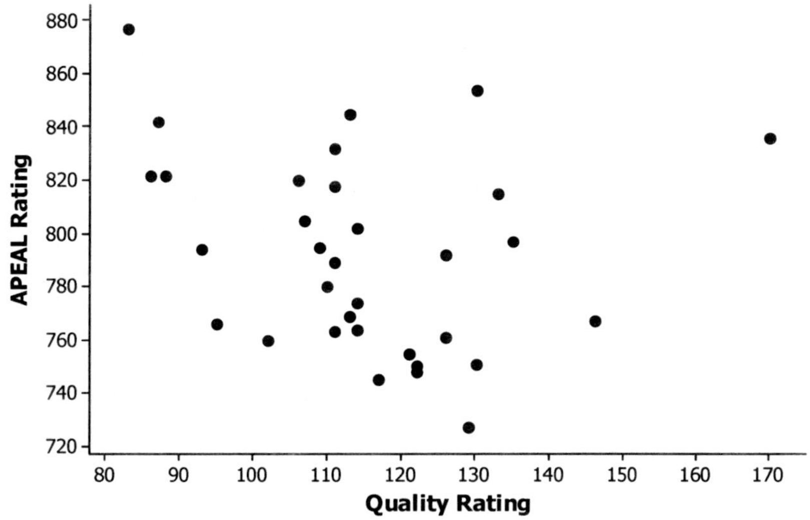

There is a weak negative relationship between customer satisfaction (as measured by the APEAL rating) and number of defects. Brands with a higher number of defects per 100 vehicles tend to have lower satisfaction ratings.

2.69: (a)

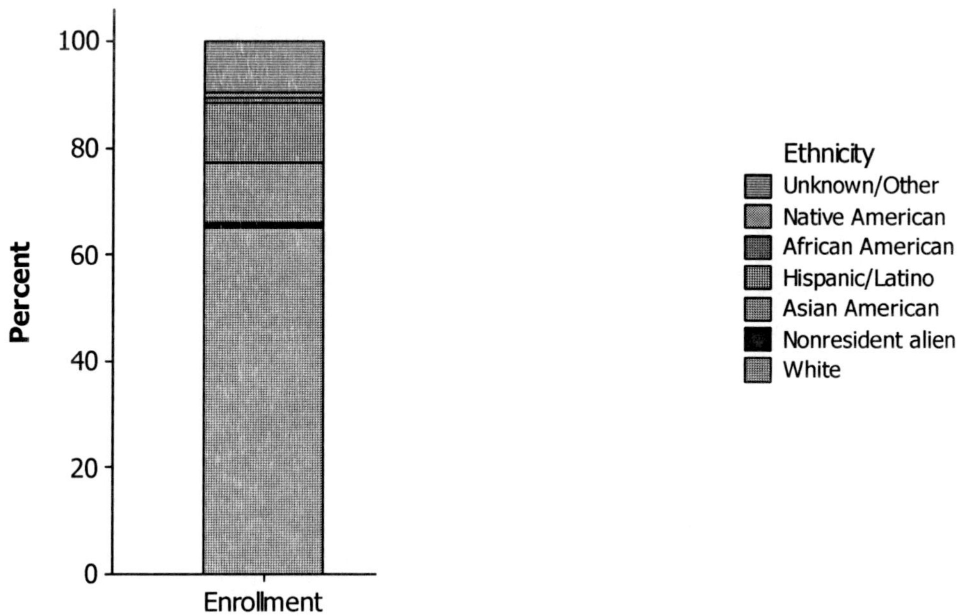

(b) The segmented bar graph in part (a) is more informative because it is easier to get a sense of the percentages of each Ethnicity who are enrolled. Specifically, in the original graphical display with the "Nonwhite" category further subdivided, it could be difficult to compare the "Nonwhite" breakout categories with the categories remaining in the pie.

(c) The pie chart combined with the segmented bar graph could have been chosen because some of the pie slices might be very thin and hard to see, and too many pieces could be difficult to visually process. In addition, it could be that the enrollment census form had the category "Nonwhite," and then that could have been further subdivided.

2.71: The display is misleading because the area principle is violated. The areas of the cocaine mounds are not proportional to the relative frequencies being represented.

Chapter 3 Numerical Methods for Describing Data Distributions

Section 3.1 Exercise Set 1

3.1: The distribution is approximately symmetric with no outliers, so the mean and standard deviation should be used to describe the center and spread, respectively.

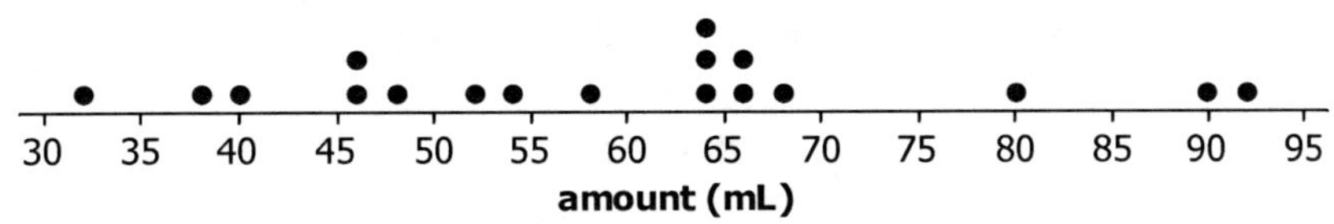

3.2: The distribution is positively skewed with an outlier, so the median and interquartile range should be used to describe the center and spread, respectively.

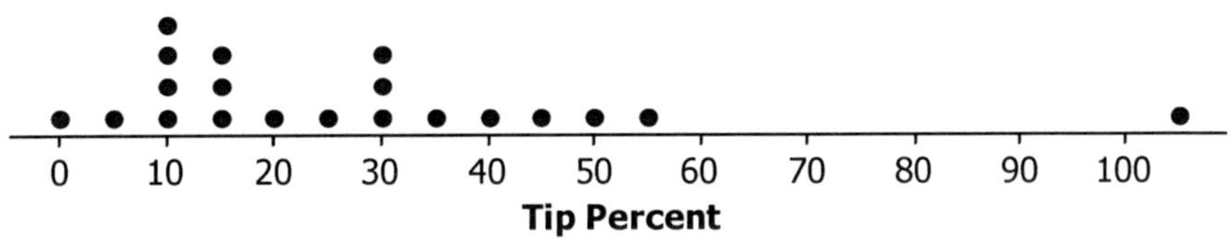

3.3: The distribution is positively skewed with a possible outlier, so the median and interquartile range should be used to describe center and spread, respectively.

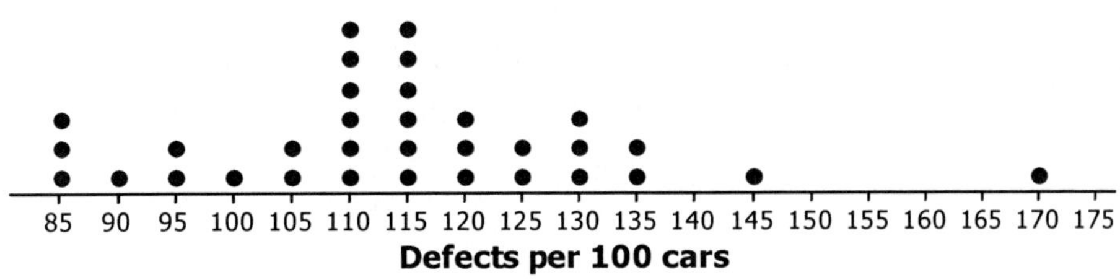

3.4: The average may not be the best measure of a typical value for this data set because examination of the dotplot (reproduced below) indicates that the distribution is clearly skewed and may contain an outlier.

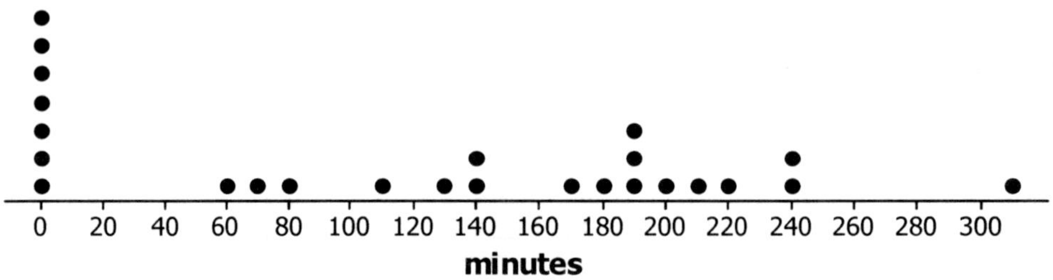

Additional Exercises for Section 3.1

3.9: The distribution is skewed, so median and interquartile range should be used to describe center and spread, respectively.

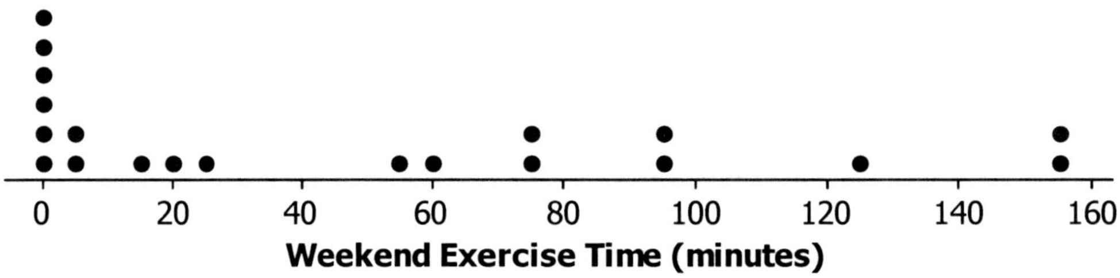

3.11: The distribution is roughly symmetric with no obvious outliers, so the mean and standard deviation should be used to describe center and spread, respectively.

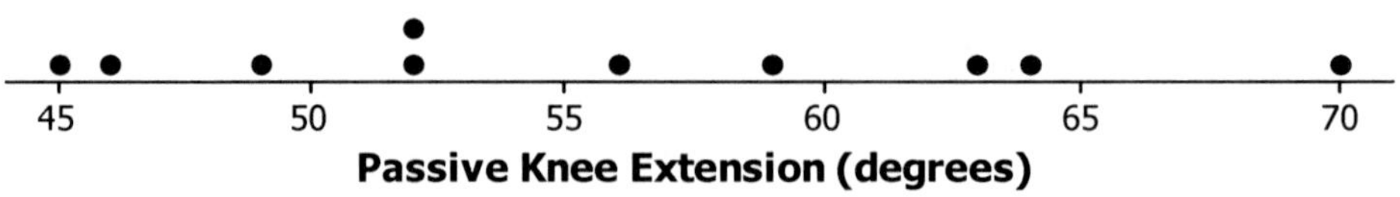

Section 3.2 Exercise Set 1

3.12: The mean is $\overline{x} = 51.33$ ounces. This is a typical or representative value for the amount of alcohol poured. The standard deviation is $s = 15.22$ ounces, which represents how much, on average, the values in the data set spread out, or deviate, from the mean.

3.13: (a) $\overline{x} = 59.23$ ounces, and $s = 16.71$ ounces. The mean represents a typical or representative value for the amount of alcohol poured and the standard deviation represents how much, on average, the values in the data set spread out, or deviate, from the mean. (b) Individuals pouring alcohol into short wide glasses pour, on average, more alcohol when pouring one shot than when pouring into tall, slender glasses.

3.14: (a) $\overline{x} = 59.85$ hours, $s = 14.78$ hours (b) $\overline{x} = 56.67$ hours, $s = 9.75$ hours. When Los Angeles was excluded from the data set, the mean and standard deviation both decreased. This suggests that using the mean and standard deviation as measures of center and spread for data sets with outliers present can be risky, because outliers seem to have a significant impact on those measures.

3.15: Answers will vary, here is one possible answer. The mean, \$444, is large, and we can likely assume that some parents spend amounts close to zero. Thus it is likely that the amounts vary greatly, making the standard deviation large.

Additional Exercises for Section 3.2

3.21: (a) $\overline{x} = 287.714$; the table below shows the deviations from the mean (b) the table below shows the sum of the deviations (at the bottom of column 2)

Data Values (x_i)	Deviation from mean $(x_i - \overline{x})$	Squared Deviations $(x_i - \overline{x})^2$
497	497 – 287.714 = 209.286	209.286^2 = 43,800.630
193	193 – 287.714 = -94.714	$(-94.714)^2$ = 8970.742
328	328 – 287.714 = 40.286	40.286^2 = 1622.962
155	155 – 287.714 = -132.714	$(-132.714)^2$ = 17,613.006
326	326 – 287.714 = 38.286	38.286^2 = 1465.818
245	245 – 287.714 = -42.714	$(-42.714)^2$ = 1824.486
270	270 – 287.714 = -17.714	$(-17.714)^2$ = 313.786
	$\sum(x_i - \overline{x}) = 0.002$	$\sum(x_i - \overline{x})^2 = 75{,}611.43$

(c) To calculate the variance and standard deviation, the squared deviations and sum of the squared deviations are needed. The third column contains these values. The variance is computed using the formula $s^2 = \frac{\sum(x_i - \bar{x})}{n-1} = \frac{75,611.43}{7-1} = 12,601.905$.

The standard deviation is $s = \sqrt{s^2} = \sqrt{12,601.905} = 112.258$.

3.23: The mean found after subtracting 10 from each sample observation is $\bar{x} = 38.36$ cm. The table below shows the original sample observations, the values after subtracting 10, and the deviations from the new mean.

Original Sample Observation	Sample Observation minus 10	Deviation from the new mean
62	52	13.64
23	13	-25.36
27	17	-21.36
56	46	7.64
52	42	3.64
34	24	-14.36
42	32	-6.36
40	30	-8.36
68	58	19.64
45	35	-3.36
83	73	34.64

The deviations for the data set obtained by subtracting 10 from each sample observation are exactly the same as the corresponding deviations from the mean for the original data set. Since the deviations are the same, the new variance (s^2) and standard deviation (s) are also the same as old variance and standard deviation. Subtracting or adding the same number to every value in a data set does not change the variance (s^2) or standard deviation (s).

Section 3.3 Exercise Set 1

3.25: (a) There is an even number of observations ($n = 20$), so the median is the average of the two middle values: $median = \frac{438,722 + 427,771}{2} = 433,246.5$. This value, 433,246.5, is the value that divides the ordered data set into two halves. This tells us that half of the values in our data set had average weekly circulations of less than 433,246.5, and the other half had average weekly circulations of more than 433,246.5. (b) The median is preferable to the mean for describing the center for this data set because the distribution is positively skewed and contains outliers. (c) It is not reasonable to generalize from this sample to the population of daily newspapers in the United States because these newspapers were not randomly selected. Rather, they are the top 20 newspapers in average weekday circulation.

3.26: Lower quartile = 10,478; upper quartile = 11,778. The lower quartile of 10,478 mg/kg is the value such that 25% of the catsups have sodium contents lower than this value, and 75% are higher. The upper quartile of 11,778 mg/kg is the value such that 75% of the catsups have sodium contents lower than this value, and 25% are higher. The interquartile range is $iqr = 11,778 - 10,478 = 1300$. The interquartile range of 1300 mg/kg is the range of the middle 50% of the catsup sodium contents. It tells us how spread out the middle 50% of the data values are.

3.27: Because $n = 25$, the median is the value in the middle of the ordered list. Therefore, the median is 142. The lower quartile is 0, and the upper quartile is 195. The interquartile range is $iqr = 195 - 0 = 195$. Half of the values of number of minutes used in cell phone calls in one month are less than or equal to 142 minutes, and half of the data values of number of minutes used in cell phone calls is greater than or equal to 142 minutes. The middle 50% of the data values have a range of 195 minutes.

3.28: The median tipping percentage is 21%. The lower quartile is 10.75%, and the upper quartile is 35.6%. The interquartile range is 35.6 – 10.75 = 24.85%. The median tipping percentage of 21% indicates that half of the tips were below 21%, and the remaining half were above 21%. The interquartile range indicates that the middle 50% of tips had a range of 24.85%.

Additional Exercises for Section 3.3

3.33: The large difference between the mean and median indicates that there were some parents who spent large amounts of money on school supplies, while the amounts for the lowest spenders were less far from the median value. These outliers have the effect of pulling the mean toward the outliers, yet the median generally remains unchanged.

3.35: (a) The dotplot is relatively symmetric, with a possible outlier at the high end of the scale. As such, the mean and median will be relatively close to each other, with the mean being greater than the median.

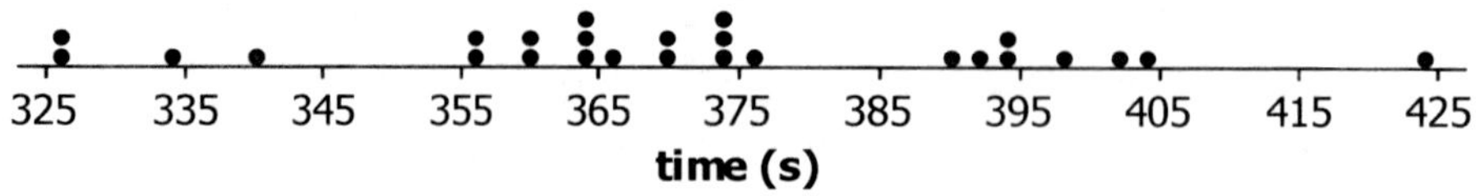

(b) $\bar{x} = 370.69$ seconds, $median = \dfrac{369+370}{2} = 369.5$ seconds. (c) The largest time could be increased by any amount and not affect the sample median because the position of the middle value will not change if the largest value is increased. The largest time could be decreased to 370 seconds without changing the value of the median.

Section 3.4 Exercise Set 1

3.37:

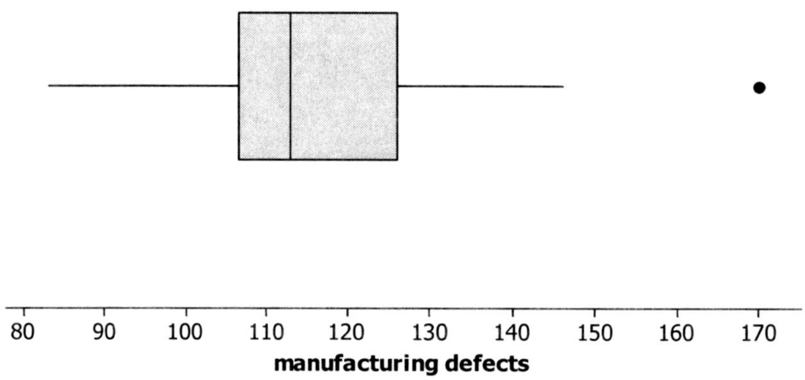

The boxplot shows that there is one outlier (170 defects), and the value of the largest non-outlier is 146 defects. The middle 50% of the data values range between about 106 and 126 defects. The distribution is positively skewed. The median is not centered in the middle 50% of the data values, further indicating the skewed nature of the distribution.

3.38: (a) No, they are not outliers. Any values greater than upper quartile + 1.5(*iqr*) or less than lower quartile – 1.5(*iqr*) are considered outliers. For this data set, values are outliers if they are greater than $32.3+1.5(32.3-20)=50.75$ cents or less than $20-1.5(32.3-20)=1.55$ cents. The largest value is not greater than 50.75, and the smallest value is not less than 1.55, so these values are not outliers.

(b)

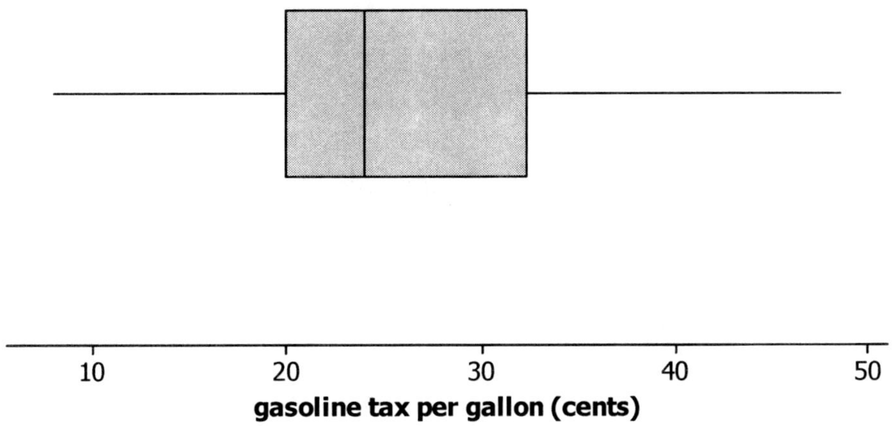

The boxplot is positively skewed. The median is not located at the center of the middle 50%, further indicating a skewed distribution.

3.39: (a) lower quartile = 16.05 inches, upper quartile = 21.93 inches.
$iqr = 21.93 - 16.05 = 5.88$ inches.

(b) Any values greater than upper quartile + 1.5(*iqr*) or less than lower quartile – 1.5(*iqr*) are considered outliers. For this data set, values are outliers if they are greater than $21.93 + 1.5(5.88) = 30.75$ inches or less than $16.05 - 1.5(5.88) = 7.23$ inches. The value 31.57 inches is an outlier.

(c)

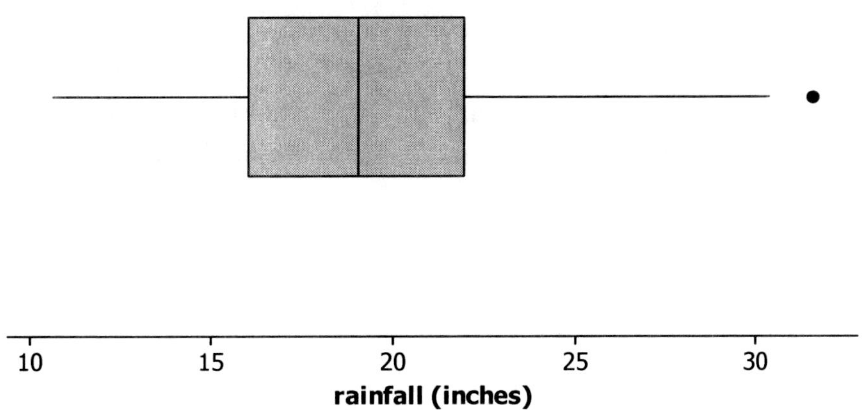

The modified boxplot shows one outlier at the high end of the scale. The distribution of inches of rainfall is slightly positively skewed.

3.40:

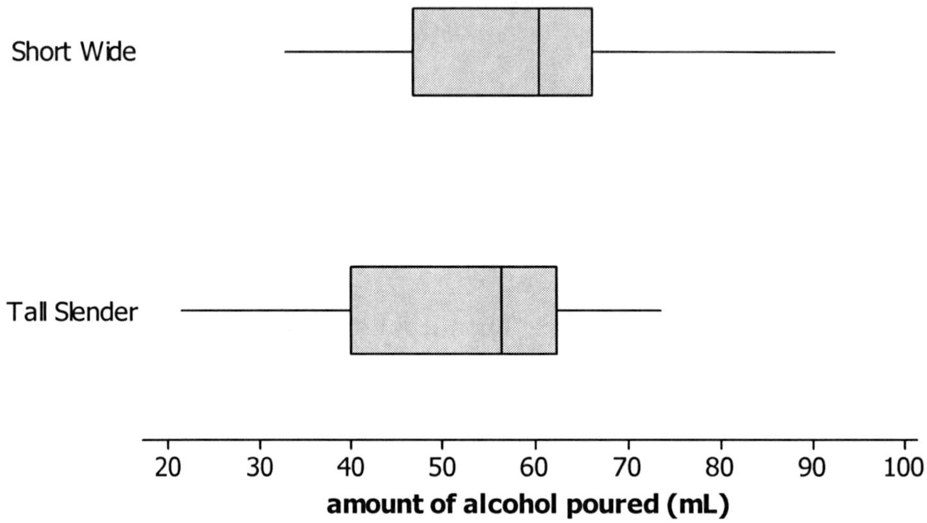

Both distributions (short wide and tall slender) are skewed, although the direction of skew is different for the two distributions. The distribution of amount of alcohol poured into short wide glasses is positively skewed, and the distribution of amount of alcohol poured into tall slender glasses is negatively skewed. The amount of alcohol poured into short wide glasses tends to be more than the amount poured into tall slender glasses. Specifically, the five-number-summary values for short wide glasses are all greater than the corresponding values for tall slender glasses. For example, the maximum amount of alcohol poured into short wide glasses (92.4 mL) is much greater than the maximum amount of alcohol poured into tall wide glasses (73.5 mL). In addition, the median amount of alcohol poured into short wide glasses (60.4 mL) is greater than the median amount of alcohol poured into tall slender glasses.

Additional Exercises for Section 3.4

3.47: (a)

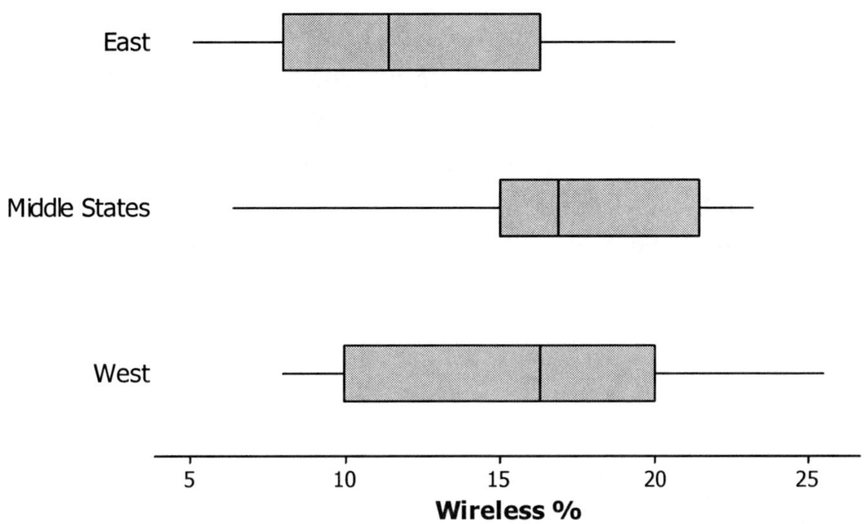

(b) The most noticeable difference between the wireless percent for the three geographical regions is that the Middle States region is negatively skewed, and has a smaller interquartile range than the East and West regions. The Eastern region has the smallest median (11.4%), and the Middle States and Western regions have medians that are much closer to each other (16.9% and 16.3%, respectively).

3.49: The fact that the mean is so much higher than the median indicates that the distribution is positively skewed. There were undoubtedly some very high punitive damage awards, which pulled the mean up toward the large values.

Section 3.5 Exercise Set 1

3.50: First national aptitude test: $z = \dfrac{625-475}{100} = 1.5$. Second national aptitude test:

$z = \dfrac{45-30}{8} = 1.875$. The student performed better on the second national aptitude test relative to the other test takers because the z-score for the second test is higher than that for the first test.

3.51: (a) 40 minutes is 1 standard deviation above the mean; 30 minutes is 1 standard deviation below the mean. The values that are 2 standard deviations away from the mean are 25 and 45 minutes. (b) Approximately 95% of times are between 25 and 45 minutes; approximately 0.3% of times are less than 20 minutes or greater than 50 minutes; Approximately 0.15% of times are less than 20 minutes.

3.52: The 10th percentile of \$0 indicates that 10% of students have \$0 or less of student debt. The 25th percentile (which is the lower quartile) indicates that 25% of students have \$0 or less of student debt. The 50th percentile (the median) indicates that 50% of students have \$11,000 or less of student debt. The 75th percentile (the upper quartile) indicates that 75% of students have \$24,600 or less of student debt. The 90th percentile indicates that 90% of students have \$39,300 or less of student debt.

3.53: (a)

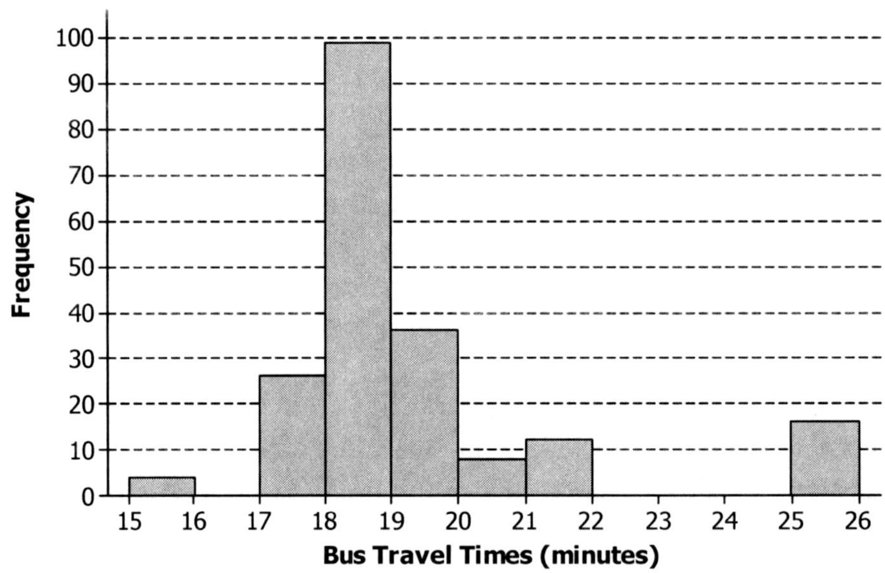

(b) (i) 86th percentile is approximately 21 minutes; (ii) 15th percentile is approximately 18 minutes; (iii) 90th percentile is approximately 21.5 minutes; (iv) 95th percentile is approximately 25.5 minutes; (v) 10th percentile is approximately 17.5 minutes

Additional Exercises for Section 3.5

3.59: (a) 1100 gallons; (b) 1400 gallons; (c) 1700 gallons

3.61: (a) 120; (b) 20; (c) $z = \frac{90-100}{20} = -0.5$; (d) 97.5; (e) Since a score of 40 is 3 standard deviations below the mean, that corresponds to a percentile of 0.15%. Therefore, there were relatively few scores below 40.

Chapter 3: Are You Ready to Move On?

3.63: The mean APEAL rating is $\bar{x} = 792.03$, which is a typical or representative value for the APEAL ratings in the sample. The standard deviation is $s = 36.70$ and represents how much, on average, the values in the data set spread out, or deviate, from the mean APEAL rating.

3.65: (a) The mean tipping percent is $\bar{x} = 27.31\%$, and the standard deviation is $s = 23.83\%$. (b) After removing the 105% tip, the new mean and standard deviation are $\bar{x}_{new} = 23.23\%$, and $s_{new} = 15.70\%$. These values are much smaller than the mean and standard deviation computed with 105 included. This suggests that the mean and standard deviation can change dramatically when outliers are present (or removed) from the data set, and, therefore, are probably not the best measures of center and spread to use in this situation.

3.67: (a) The median is 140 seconds, and the interquartile range is $iqr = 200 - 100 = 100$ seconds. The median divides the ordered list into two equal halves, with half the values less than 140 seconds and half the values greater than 140 seconds. The interquartile range of 100 seconds indicates that the middle 50% of the data values have a range of 100 seconds. (b) Due to the presence of the large outlier (the value 380 seconds), the median and interquartile range are the appropriate summary measures to describe center and spread for this data set.

3.69: (a) Median = 8 grams/serving; lower quartile = 7 grams/serving; upper quartile = 12 grams/serving; interquartile range = 12 – 7 = 5 grams/serving

(b) Median = 10 grams/serving; lower quartile = 6 grams/serving; upper quartile = 13 grams/serving; interquartile range = 13 – 6 = 7 grams/serving

(c) There are no outliers in the sugar content data set because there are no values greater than 1.5(*iqr*) above the upper quartile or smaller than 1.5(*iqr*) below the lower quartile.

(d) The minimum value and lower quartile are the same because the smallest five values in the data set are all equal to 7.

(e)

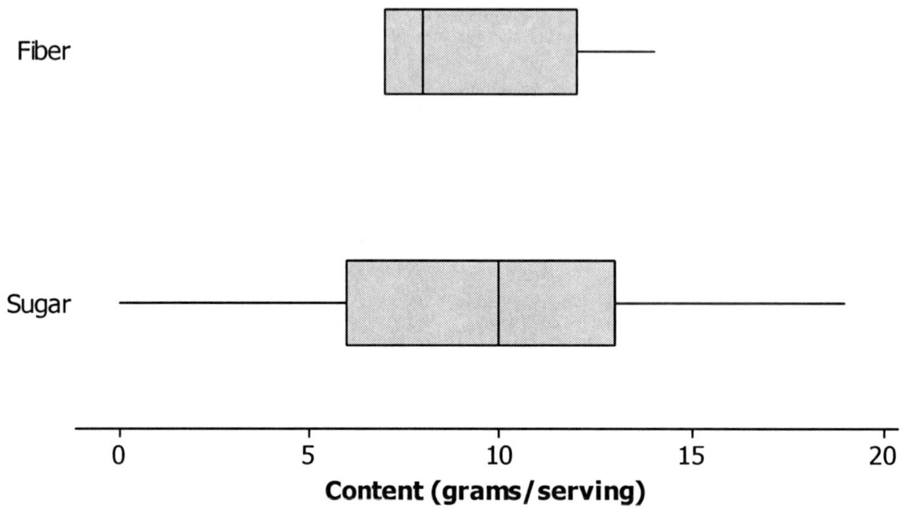

The sugar content in grams/serving is much more variable than the fiber content in grams/serving. The range in sugar content (19 grams/serving) is greater than the range in fiber content (7 grams/serving). The boxplot of fiber content shows that the minimum and lower quartile are equal to each other, which is not observed in the sugar content. The distribution of sugar content values are approximately symmetric, which is different from the skewed fiber distribution.

3.71: (a) The 25th percentile indicates that 25% of full-time female workers age 25 or older with an Associate degree earn $26,800 or less. The 50th percentile indicates that 50% of full-time female workers age 25 or older with an Associate degree earn $36,800 or less. The 75th percentile indicates that 75% of full-time female workers age 25 or older with an Associate degree earn $51,100 or less. (b) The 25th, 50th, and 75th percentile values for men are all greater than the corresponding percentiles for female workers, indicating that full-time employed men age 25 or older with an Associate degree, in general, earn more than full-time employed women age 25 or older with an Associate degree.

Chapter 4 Describing Bivariate Numerical Data

Section 4.1 Exercise Set 1

4.1: Scatterplot 1: (i) Yes, (ii) Yes, (iii) Negative

Scatterplot 2: (i) Yes, (ii) No, (iii) –

Scatterplot 3: (i) Yes, (ii) Yes, (iii) Positive

Scatterplot 4: (i) Yes, (ii) Yes, (iii) Positive

4.2: (a) Negative correlation, because as interest rates rise, the number of loan applications might decrease. (b) Close to zero, because there is no reason to believe that height and IQ should be related. (c) Positive correlation, because taller people tend to have larger feet. (d) Positive correlation, because as the minimum daily temperature increases, the cooling cost would also increase.

4.3: (a) There is a moderately strong positive association between school achievement test score and midlife IQ. (b) $r = 0.6$, because the article says that $r = 0.64$ indicated a very strong relationship, higher than the correlation between height and weight in adults. Therefore, a correlation that is moderately strong ($r = 0.6$) with a positive association (taller people tend to weigh more) is consistent with the statement.

4.4: (a) $r = 0.335$. The correlation indicates that there is a weak, positive linear relationship between number of acres burned and timber sales. (b) The conclusion that "heavier logging led to large forest fires" cannot be justified because correlation does not imply causation; the correlation only measures the strength and direction of the association between two quantitative variables.

4.5: (a)

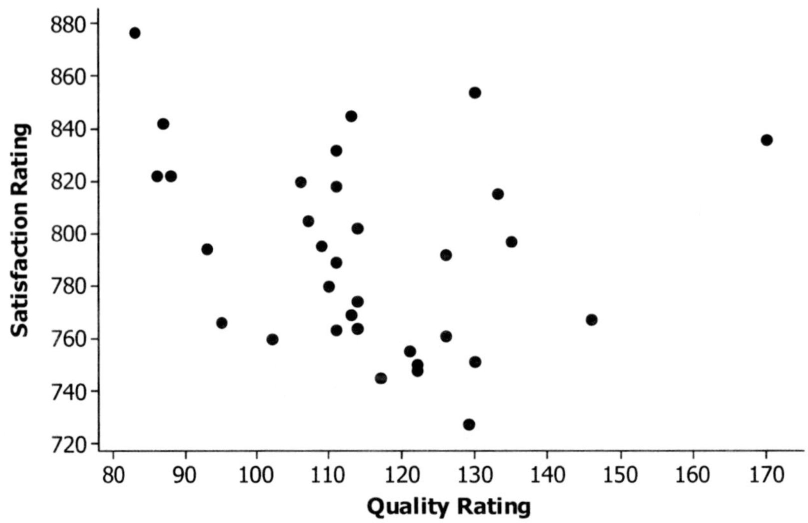

There is a weak, negative linear relationship between satisfaction rating and quality rating.

(b) The correlation of r = -0.239 indicates that there is a weak, negative linear relationship between satisfaction rating and quality rating.

4.6: No, the statement is not correct. A correlation of 0 indicates that there is not a linear relationship between two variables. There could be a strong non-linear relationship (for example, a quadratic relationship) between the two variables.

4.7: No, it is not reasonable to conclude that increasing alcohol consumption will increase income. Correlation measures the strength of association, but *association does not imply causation.* It could be that people with higher incomes can afford to drink more alcohol.

4.8: The correlation between college GPA and academic self-worth (r = 0.48) indicates that there is a weak or moderate positive linear relationship between those variables. This tells us that those athletes with higher a GPA tend to feel better about themselves academically than those with lower grades. The correlation between college GPA and high school GPA (r = 0.46) indicates that there is a weak or moderate positive relationship between those variables as well. This tells us that those athletes with higher high school GPA tend to also have a higher college GPA. Finally, the correlation between college GPA and a measure of tendency to procrastinate (r = -0.36) indicates that there is a weak negative linear relationship between those variables. Those athletes with a lower college GPA tend to procrastinate more than those athletes with a higher college GPA.

Additional Exercises for Section 4.1

4.15: (a)

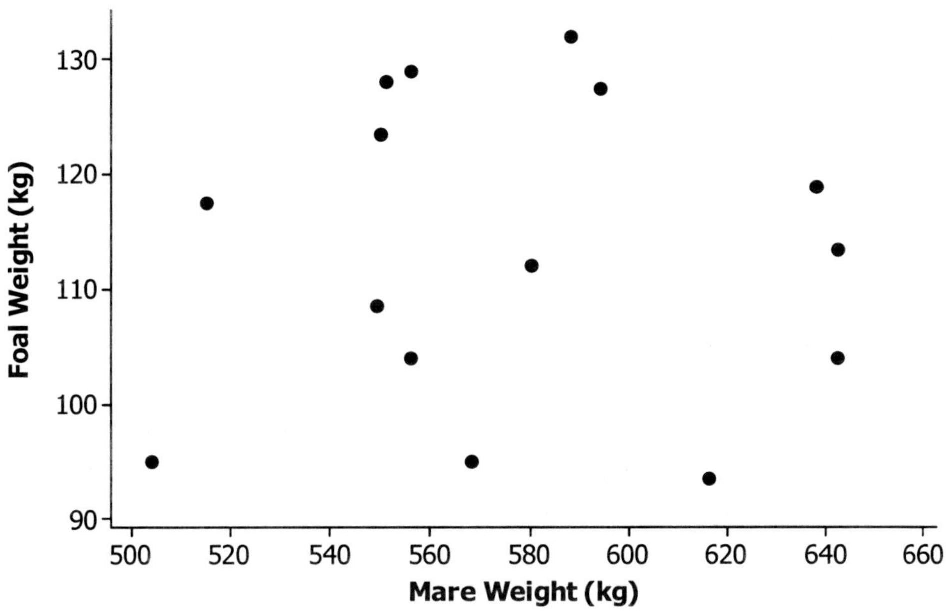

(b) $r = 0.001$ (c) The scatterplot shows that there is no obvious relationship (linear or otherwise) between mare weight and foal weight. The correlation coefficient of $r = 0.001$ indicates that there is essentially no linear relationship between mare weight and foal weight.

4.17: $r = 0.987$; this value is consistent with the previous answer, because the correlation coefficient is large and positive (close to 1), which indicates a strong positive association between household debt and corporate debt.

4.19: The sample correlation coefficient would be closest to -0.9. Cars traveling at a faster rate of speed will travel the length of the highway segment more quickly than those who are traveling more slowly, and we can expect this relationship to be strong.

Section 4.2 Exercise Set 1

4.21: It makes sense to use the least squares regression line to summarize the relationship between x and y for Scatterplot 1, but not for Scatterplot 2. Scatterplot 1 shows a linear relationship between x and y, but Scatterplot 2 shows a curved relationship between x and y.

4.22: The sum of the squared vertical deviations from the line $y = 12.5 - 0.5x$ would be larger than the sum of the squared vertical deviations from the least squares regression line $\hat{y} = 12 - 0.36x$ because, by definition, the least squares regression line is *the* line with the minimum value for the sum of the squared vertical deviations from the line. All other lines would have larger values for the sum of the squared vertical deviations.

4.23: (a) $\hat{y} = -5.0 + 0.017x$, where $\hat{y}$ represents the predicted amount of natural gas used and x represents the size of a house. (b) $\hat{y} = -5.0 + 0.017(2100) = 30.7$ therms. (c) This is the slope of the line, 0.017. (d) No, because the regression line was determined based on house sizes between 1000 and 3000 square feet. There is no guarantee that the linear relationship will continue outside this range of house sizes. This is the danger with extrapolation.

4.24: (a) The response variable (y) is birth weight, and the predictor variable (x) is mother's age.

(b) It is reasonable to use a line to summarize the relationship between birth weight and mother's age. There is a clear linear relationship between birth weight and mother's age.

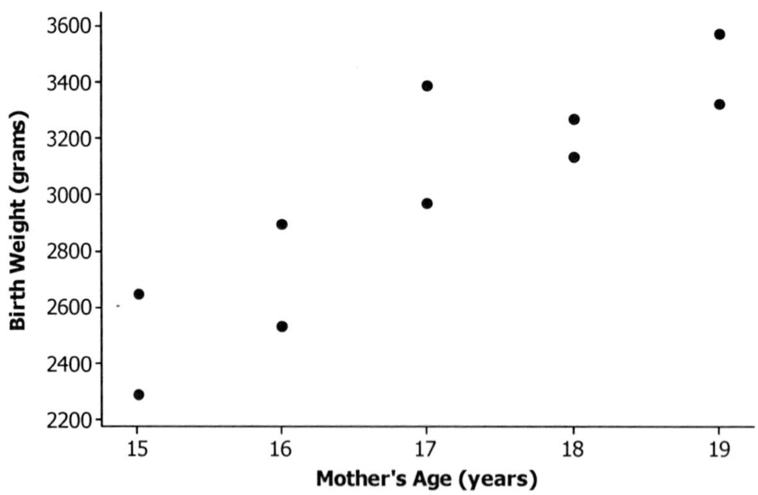

(c) $\hat{y} = -1163.4 + 245.15x$, where y represents the birth weight in grams, and x represents the mother's age in years.

(d) The slope of 245.15 is the amount, on average, by which the birth weight increases when the mother's age increases by one year.

(e) It is not appropriate to interpret the intercept of the least squares regression line. The intercept is the birth weight for a mother who is zero years old, which is impossible. In addition, the intercept is negative, indicating a negative birth weight, which is also impossible.

(f) $\hat{y} = -1163.4 + 245.15(18) = 3249.3$ grams

(g) $\hat{y} = -1163.4 + 245.15(15) = 2513.85$ grams

(h) It is not advisable to use the least squares regression line to predict the birth weight of a baby born to a 23 year old mother because 23 years is well outside the range of ages used in determining the least squares regression line. There is no guarantee that the linear relationship between birth weight and mother's age will continue outside the range of 15 to 19 years.

4.25: (a) The response variable is the cost of medical care, and the predictor variable is the measure of pollution.

(b)

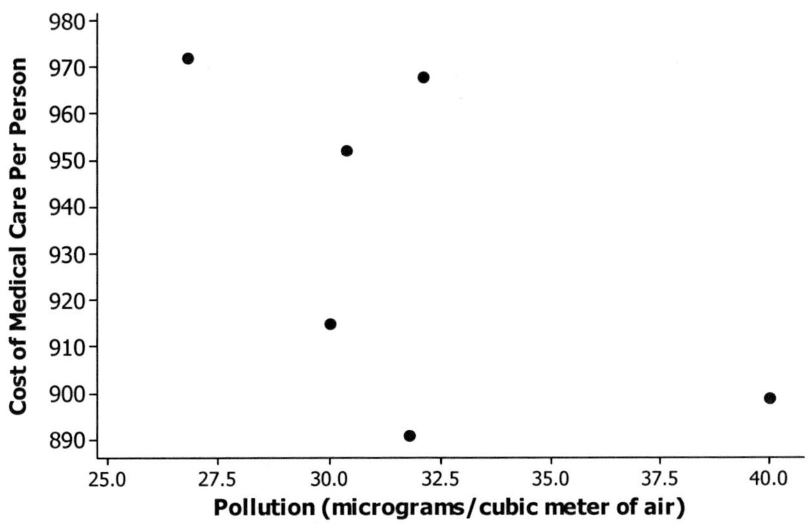

There is a moderate, negative association ($r = -0.581$) between pollution and cost of medical care per person.

(c) $\hat{y} = 1082.2 - 4.691x$, where y is the cost of medical care per person, and x is the pollution measure.

(d) The slope is negative, and it is consistent with the observed negative association in the scatterplot, as well as with the negative correlation coefficient.

(e) No, the scatterplot and least squares regression line equation do not support the researcher's conclusion that people over the age of 65 who live in more polluted areas have higher medical costs. The association between medical cost and pollution level is negative, which indicates that people over the age of 65 in more polluted areas tend to have lower medical costs.

(f) $\hat{y} = 1082.2 - 4.691(35) = \$918.02$

(g) No, I would not use the least squares regression equation to predict the cost of medical care for a region that has a pollution measure of 60. The data set that was used to create the least squares regression line was based on pollution measures that vary between 26.8 and 40; the value 60 is far outside that range of values. There is no guarantee that the observed trend will continue as far as 60. This is a danger of extrapolation.

4.26: (a) The scatterplot (shown below) shows a strong linear relationship between the air flow readings from the mini-Wright and Wright meters. The correlation coefficient ($r = 0.943$) supports the observation made from the scatterplot. Therefore, we can confidently use the data to find the equation of the least squares regression line. That equation is $\hat{y} = 11.48 + 0.970x$, where $\hat{y}$ is the predicted Wright meter reading and x is the mini-Wright meter reading.

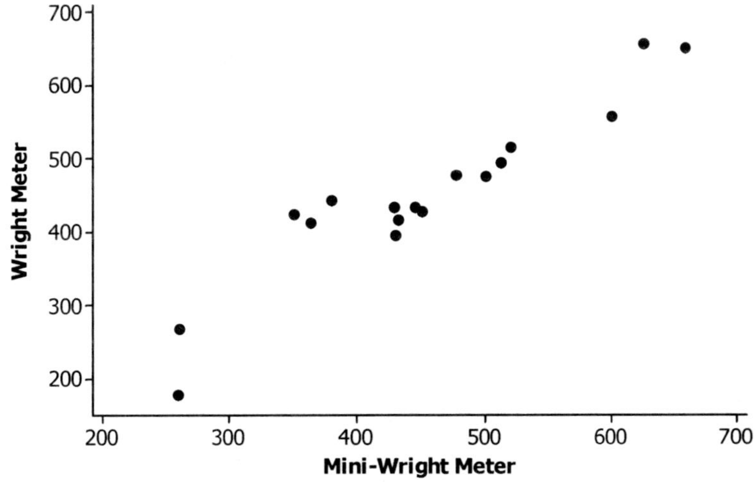

(b) The predicted Wright meter reading for a mini-Wright meter reading of 500 is $\hat{y} = 11.48 + 0.970(500) = 496.48$.

(c) The predicted Wright meter reading for a mini-Wright meter reading of 300 is $\hat{y} = 11.48 + 0.970(300) = 302.48$.

Additional Exercises for Section 4.2

4.33: $\hat{y} = 13.5 - 0.195x$, where $\hat{y}$ is the predicted number of cell phone calls, and x is the age.

4.35: Scatterplot of cell phone calls versus age:

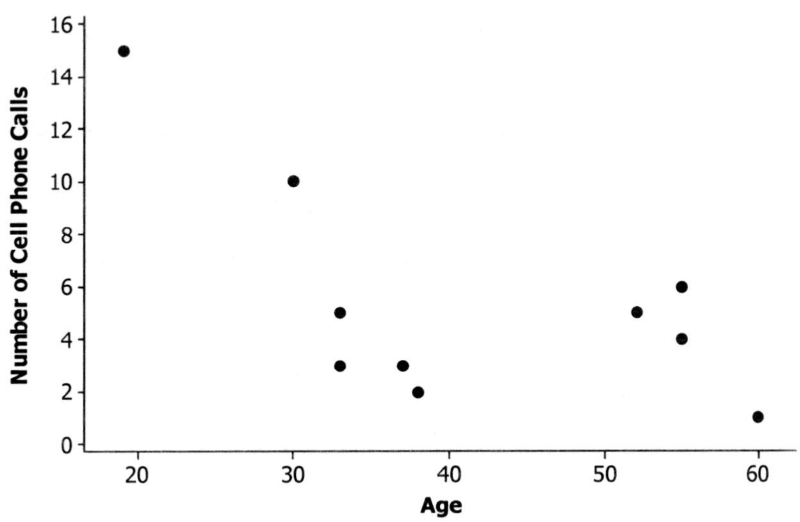

Scatterplot of text messages sent versus age:

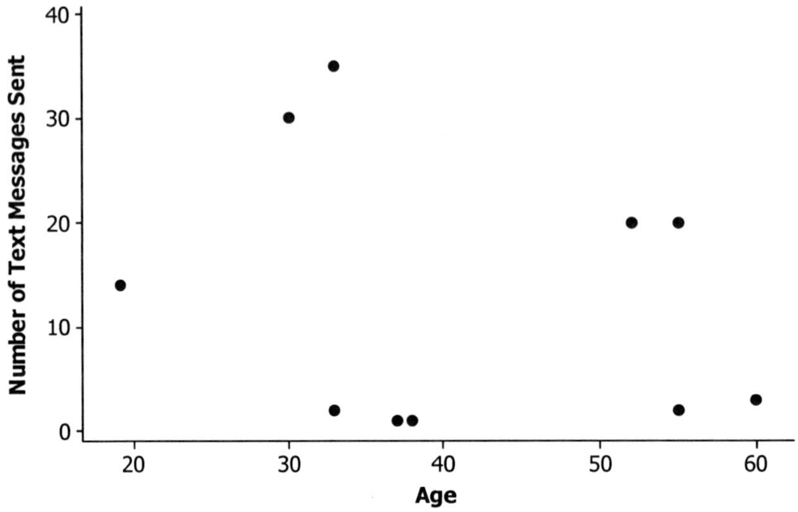

Age is a better predictor of number of cell phone calls. The linear relationship between age and number of cell phone calls is stronger than the relationship between age and number of text messages sent.

4.37: The slope is the change in predicted price for each additional mile from the Bay, so the slope would be –4,000.

Section 4.3 Exercise Set 1

4.39: (a) $\hat{y} = 1.33878 - 0.007661x$, where $\hat{y}$ is the predicted telomere length, and x is the perceived stress. (b) $r^2 = 0.099$; Approximately 9.9% of the variability in telomere length can be explained by the linear relationship between telomere length and perceived stress. (c) $s_e = 0.159119$ (obtained using software); s_e tells us that telomere length will deviate from the least squares regression line by 0.159119, on average. (d) The relationship between perceived stress and telomere length is weak and negative. It is negative because the slope of the least squares regression line is negative, and it is weak because the correlation coefficient (which has the same sign as the slope) is $r = -\sqrt{r^2} = -\sqrt{0.099} = -0.315$.

4.40: A small value of s_e indicates that residuals tend to be small. Because residuals represent the difference between an observed y value and a predicted y value, the value of s_e tells us how much accuracy we can expect when using the least squares regression line to make predictions.

4.41: It is important to consider both r^2 and s_e when evaluating the usefulness of the least squares regression line because a large r^2 (which indicates the proportion of variability in y that can be explained by the linear relationship between x and y) tells us that knowing the value of x is helpful in predicting y, and a small s_e indicates that residuals tend to be small, thus the data points tend to be relatively close to the regression line. These are both desirable when assessing the usefulness of the regression line.

4.42: (a) Yes, the scatterplot looks reasonably linear.

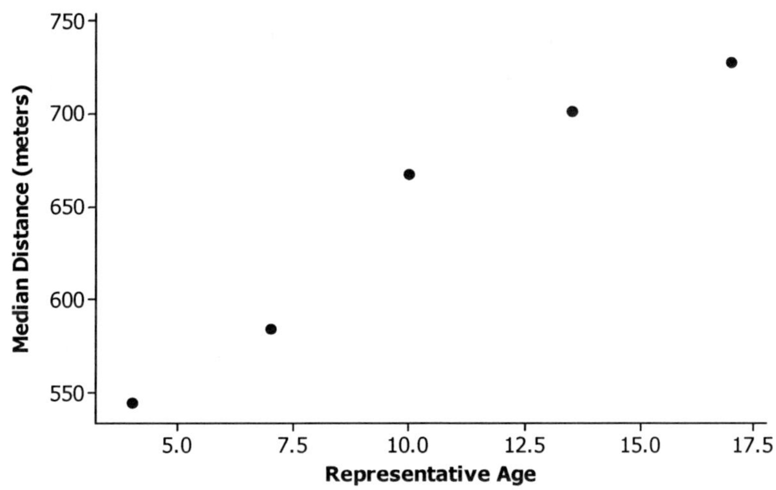

(b) $\hat{y} = 492.80 + 14.763x$, where $\hat{y}$ is the predicted median distance walked, and x is the representative age.

(c) The residuals are shown in the last column of the table below. The predicted y ($\hat{y}$) values were found using the least squares regression equation.

Representative Age (x)	Median Distance Walked (y)	Predicted y ($\hat{y}$)	Residual ($y - \hat{y}$)
4.0	544.3	551.851	-7.5510
7.0	584.0	596.141	-12.1410
10.0	667.3	640.431	26.8690
13.5	701.1	692.103	8.9974
17.0	727.6	743.774	-16.1743

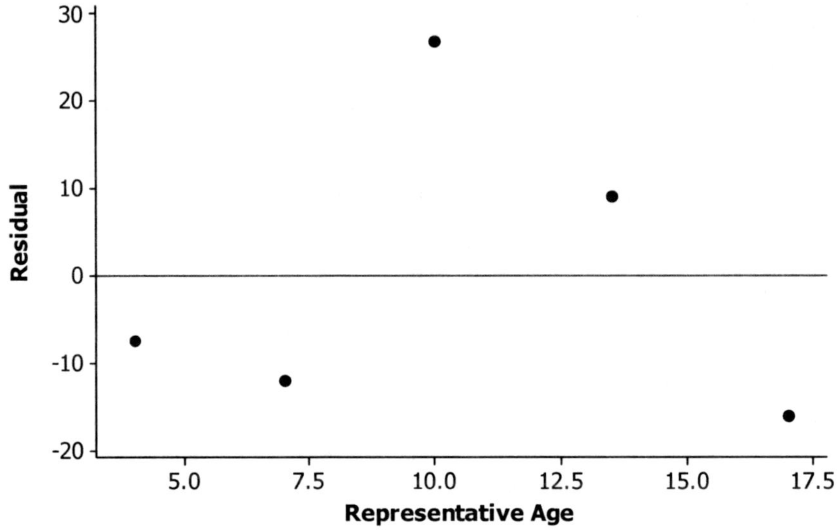

Yes, there is an unusual feature in the plot, namely, there is a curved pattern in the residual plot. The curvature indicates that the relationship between median distance walked and representative age is not linear.

4.43: (a)

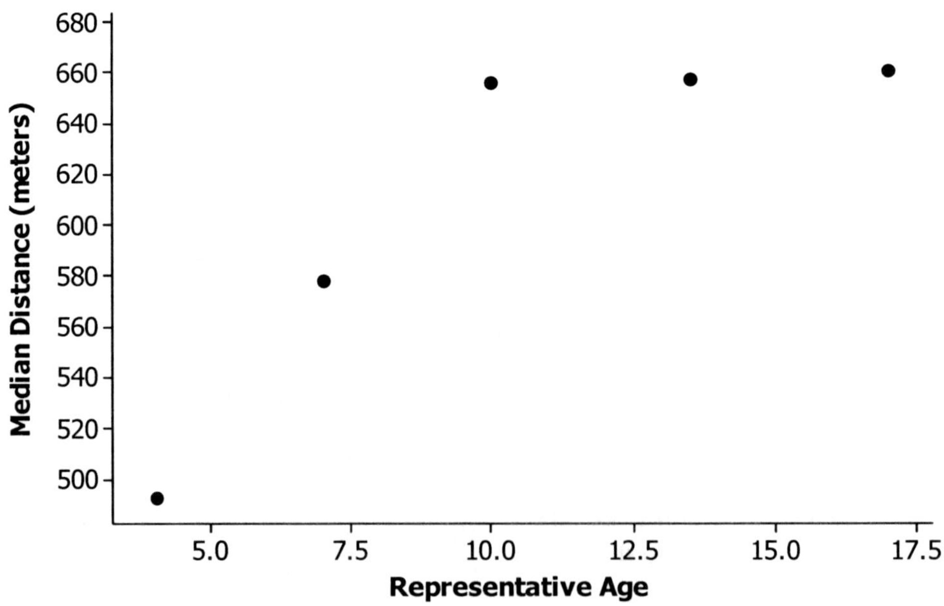

The pattern for girls differs from boys in that the girls' scatterplot shows more apparent nonlinearity in the pattern.

(b) $\hat{y} = 480 + 12.525x$, where $\hat{y}$ is the girls' predicted distance walked, and x is the representative age.

(c)

Representative Age (x)	Median Distance Walked (y)	Predicted y ($\hat{y}$)	Residual ($y - \hat{y}$)
4.0	492.4	530.095	-37.6952
7.0	578.3	567.669	10.6311
10.0	655.8	605.243	50.5574
13.5	657.6	649.079	8.5214
17.0	660.9	692.915	-32.0147

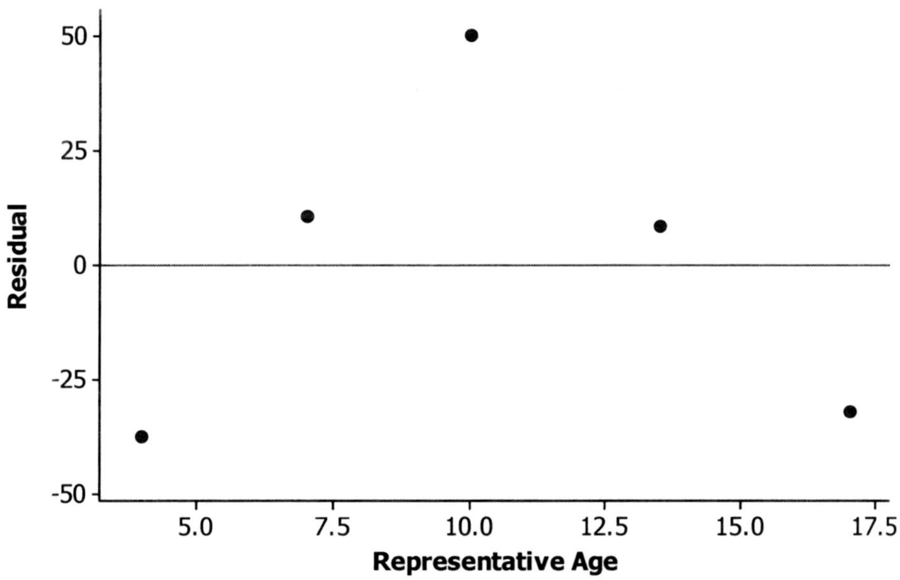

The curvature of the residual plot indicates that a curve is more appropriate than a line for describing the relationship between median distance walked and representative age.

4.44: (a)

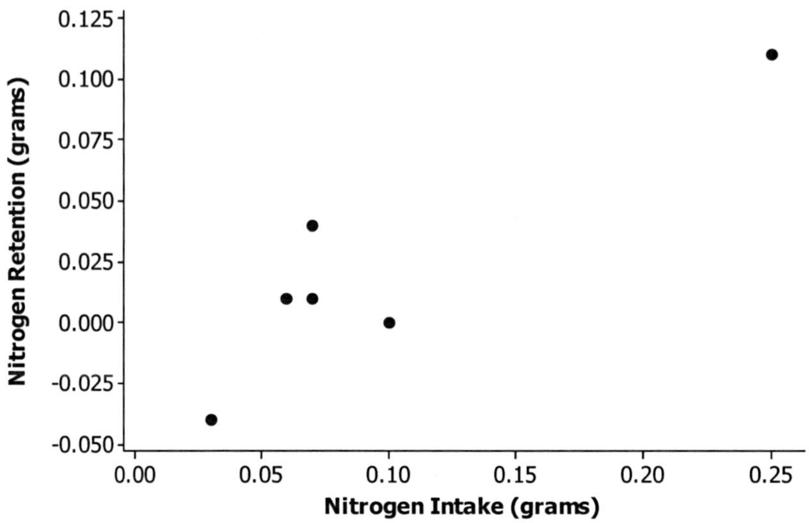

(b) $\hat{y} = -0.03443 + 0.5803x$, where $\hat{y}$ is the predicted nitrogen retention, and x is the nitrogen intake (both in grams). The predicted nitrogen retention for a flying squirrel whose nitrogen intake is 0.06 grams is $\hat{y} = -0.03443 + 0.5803(0.06) = 0.000388$ grams. The residual associated with the observation (0.06, 0.01) is $y - \hat{y} = 0.01 - 0.000388 = 0.009612$.

(c) The observation (0.25, 0.11) is potentially influential because that point has an x value that is far away from the rest of the data set.

(d) $\hat{y} = -0.037 + 0.627(0.06) = 0.00062$; This prediction is larger than the prediction made in part (b).

4.45: (a)

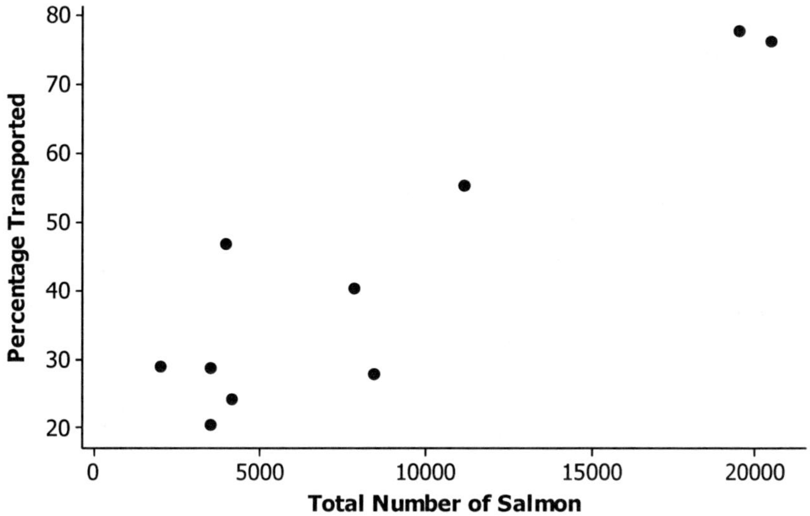

There appears to be a linear relationship between Percentage Transported and Total Number of Salmon.

(b) $\hat{y} = 18.483 + 0.0028655x$, where $\hat{y}$ is the predicted percentage transported and x is the total number of salmon.

(c) The observation (3928, 46.8) is not influential, because the x-value for that observation is not far from the rest of the data. In addition, removal of the potentially influential point produces a least squares regression line with a y-intercept and slope similar to the original line.

(d) Those points are not considered influential even though they are far from the rest of the data because they follow the trend of the remaining data points. Removal of those potentially influential points would produce a least squares regression line similar to the line found using the full data set.

(e) $s_e = 9.16217$; Observations tend to deviate from the least squares regression line by 9.16217 percentage points on average.

(f) $r^2 = 0.832$; Approximately 83.2% of the variability in percentage transported can be explained by the linear relationship between percentage transported and number of salmon.

Are You Ready to Move On? Chapter 4 Review Exercises

4.57: Scatterplot 1: (i) Yes, (ii) Yes, (iii) Negative

Scatterplot 2: (i) Yes, (ii) No, (iii) –

Scatterplot 3: (i) Yes, (ii) Yes, (iii) Positive

Scatterplot 4: (i) Yes, (ii) Yes, (iii) Positive

4.59: (a) r = -0.10; there is a weak, negative linear relationship. Because the relationship is negative, larger arch heights tend to be paired with smaller average hopping heights.

(b) The correlation coefficients support the conclusion since they are all fairly close to 0.

4.61: (a) r = 0.944; there is a strong positive linear relationship between sugar consumption and depression rate.

(b) No, because you can't conclude that a cause-and-effect relationship exists just based on a strong correlation.

(c) These countries were not a random sample of all countries, and it is unlikely that they are representative.

4.63: (a) Negative, because it is likely that the work is more stressful and less enjoyable for nurses with high patient-to-nurse ratios.

(b) Negative, because it is likely that the patients at hospitals where the patient-to-nurse ratio is high will not get as much individual attention and will be less satisfied with their care.

(c) Negative, because it is likely that the quality of patient care suffers when patient-to-nurse ratio is high.

4.65: There is no evidence that the form of the relationship between x and y remains the same outside the range of the data.

4.67: (a) Size, because the relationship between price and size (r = 0.700) is stronger than the relationship between price and land-to-building ratio (r = -0.332).

(b) $\hat{y} = 1.33 + 0.00525x$, where $\hat{y}$ is the predicted sale price and x is the size in thousands of square feet.

4.69: (a)

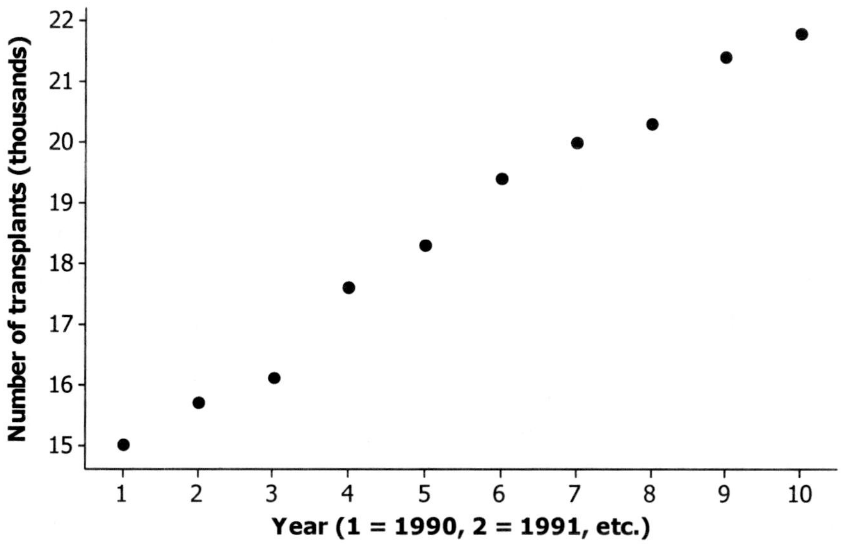

$\hat{y} = 14.2 + 0.790x$, where $\hat{y}$ is the predicted number of transplants and x is the year (using 1, 2, …, 10 to represent the years). The number of transplants has increased steadily over time, with a predicted increase of about 790 per year.

(b)

Year (x)	Number of Transplants (y)	Predicted Number of Transplants ($\hat{y}$)	Residuals ($y - \hat{y}$)
1	15.0	15.0036	-0.003636
2	15.7	15.7939	-0.093939
3	16.1	16.5842	-0.484242
4	17.6	17.3745	0.225455
5	18.3	18.1648	0.135152
6	19.4	18.9552	0.444848
7	20.0	19.7455	0.254545
8	20.3	20.5358	-0.235758
9	21.4	21.3261	0.073939
10	21.8	22.1164	-0.316364

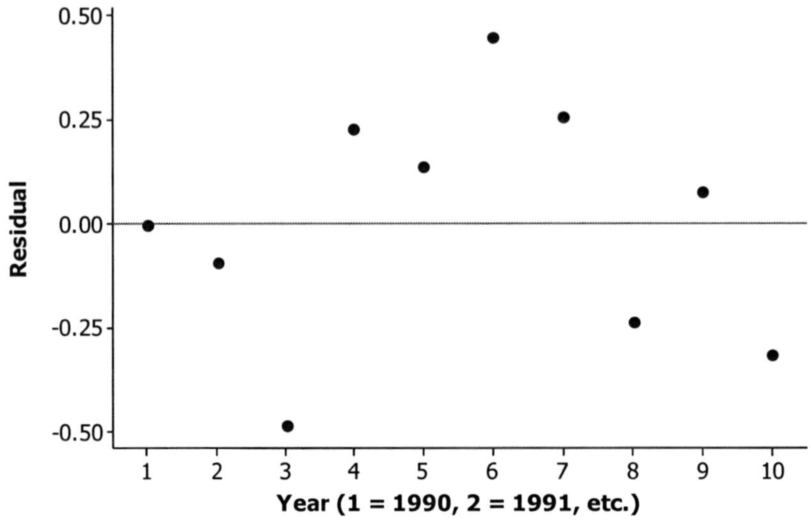

The residual plot shows a curved pattern, suggesting that the relationship between year and number of transplants is nonlinear.

4.71: (a) No, because the value of outpatient cost-to-charge ratio (54) is not far from the other outpatient values in the data set.

(b) Yes, it is an outlier, because it has a large residual—it is far away from the least-squares regression line.

(c) No, because even though it has an outpatient value that is larger than the others in the data set, this data point is consistent with the linear pattern formed by the other data value.

(d) No, because it falls quite close to the least squares regression line. This data point has a small residual.

Chapter 5 Probability

Section 5.1 Exercise Set 1

5.1: (a) In the long run, about 1% of people who suffer cardiac arrest in New York City will survive. (b) Because 1% of 2329 is 23.29, approximately 23 people who suffer cardiac arrest in New York City survived.

5.2: In the long run, 86% of the time this particular flight that flies between Phoenix and Atlanta will arrive on time.

5.3: (a) P(subscriber downloads three or fewer movies) = 0.03 + 0.45 + 0.25 + 0.10 = 0.83

(b) P(subscriber downloads at most three movies) = 0.03 + 0.45 + 0.25 + 0.10 = 0.83

(c) P(subscriber downloads four or more movies) = 0.10 + 0.07 = 0.17

(d) P(subscriber downloads zero or one movie) = 0.03 + 0.45 = 0.48

(e) P(subscriber downloads more than one movie) = 0.25 + 0.10 + 0.10 + 0.07 = 0.52

Additional Exercises for Section 5.1

5.7: Approximately 167; P(observing six when a fair die is rolled) = 1/6

5.9: (a) P(purchased all print books online) = 0.17

(b) P(purchased all print books) = 0.25 + 0.12 + 0.17 + 0.31 = 0.85

(c) P(purchased some e-books or all e-books) = 0.04 + 0.11 = 0.15

(d) P(did not purchase all print books from the campus bookstore) = 1 – 0.25 = 0.75

Section 5.2 Exercise Set 1

5.11: (a) $P(\text{Phoenix as final destination}) = \dfrac{1800}{8000} = 0.225$

(b) $P(\text{did not have Phoenix as final destination}) = \dfrac{6200}{8000} = 0.775$

(c) $P(\text{was connecting and missed connecting flight}) = \dfrac{480}{8000} = 0.06$

(d) $P(\text{was connecting and did not miss connecting flight}) = \dfrac{5720}{8000} = 0.715$

(e) $P(\text{Phoenix as final destination or was delayed overnight}) = \dfrac{1875}{8000} = 0.234$

(f) 75/8000 is the probability, for any person in the sample, that that person was delayed overnight. We would expect the number of people in the sample who were delayed overnight to be around 50(75/8000) = 0.469. This is less than one person, and so the airline should not be worried about having a large number of people who were delayed overnight being included in the sample.

5.12: (a) The sample space is $S = \{AB, AC, AD, AE, BC, BD, BE, CD, CE, DE\}$

(b) Yes, the outcomes are equally likely because the problems are being chosen at random.

(c) You did only three of the problems, and there are exactly three ways to select two of the three problems you did. Because there are ten possible pairs of problems that could be chosen, the probability that you will be able to turn in both of the problems selected is 3/10 = 0.3.

(d) No, the probability does not change. The problems are being selected at random, and each problem is equally likely. No matter what three problems you completed, there are still exactly three out of the possible ten pairs that would allow you to turn in both problems.

(e) The probability increases to 6/10 = 0.6.

5.13: $P(\text{the patient develops an infection}) \approx \dfrac{712}{6450} = 0.11$

Additional Exercises for Section 5.2

5.17: (a) $P(\text{the selected student will be female}) = \dfrac{488{,}142}{787{,}325} = 0.62$

(b) $P(\text{the selected student will be male}) = \dfrac{299{,}183}{787{,}325} = 0.38$

5.19: (a) $P(\text{randomly selected pregnant woman delivers twins}) = \dfrac{500{,}000}{42{,}005{,}100} = 0.0119$

(b)

$P(\text{randomly selected pregnant woman delivers quadruplets}) = \dfrac{100}{42{,}005{,}100} = 0.000002$

(c)

$P(\text{randomly selected woman gives birth to more than a single child}) = \dfrac{505{,}100}{42{,}005{,}100} = 0.012$

5.21: (a) $S = \{AB, AC, AD, AE, AF, BC, BD, BE, BF, CD, CE, CF, DE, DF, EF\}$

(b) Yes, the outcomes are equally likely.

(c) $P(\text{both contestants are under 25}) = \dfrac{1}{15} = 0.0667$

(d) $P(\text{both the chosen contestants are not younger than 25}) = \dfrac{6}{15} = 0.4$

(e) $P(\text{one is younger than 25 and the other is not}) = \dfrac{8}{15} = 0.533$

Section 5.3 Exercise Set 1

5.22: (a) Not mutually exclusive, because there are seniors who are computer science majors.

(b) Not mutually exclusive, because there are female students who are computer science majors.

(c) Mutually exclusive, because a student cannot both have a college residence that is more than 10 miles from campus and live in a college dormitory that is on campus.

(d) Mutually exclusive, because college football teams are male teams and females would not be on the team.

5.23: The events are dependent because the probability that a male experiences pain daily is not equal to the probability that a female experiences pain daily.

5.24: (a) $P(\text{selected loan will be for an adjustable rate mortgage}) = \frac{2625}{3750} = 0.70$

(b) $P(\text{selected loan will be for a multifamily property}) = \frac{710}{3750} = 0.189$

(c) $P(\text{selected loan will not be for a single} - \text{family property}) = \frac{1165 + 710}{3750} = 0.5$

(d)

$P(\text{selected loan will be for a single} - \text{family property or a condo}) = \frac{1875 + 1165}{3750} = 0.811$

(e)

$P(\text{multifamily property or for an adjustable rate loan}) = \frac{337 + 373 + 1500 + 788}{3750} = 0.799$

(f) $P(\text{selected loan will be a fixed} - \text{rate loan for a condo}) = \frac{377}{3750} = 0.101$

5.25: (a)

	E **(stop at first light)**	***not E*** **(does not stop at first light)**	**Total**
F **(stop at second light)**	150	150	300
not F **(does not stop at second light)**	250	450	700
Total	400	600	1000

(b) (i) $P(\text{Shelly must stop for at least one light}) = \frac{250 + 150 + 150}{1000} = 0.55$

(ii) $P(\text{Shelly does not have to stop at either light}) = \frac{450}{1000} = 0.45$

(iii) $P(\text{Shelly must stop at exactly one of the two lights}) = \dfrac{150+250}{1000} = 0.40$

(iv) $P(\text{Shelly must stop only at the first light}) = \dfrac{250}{1000} = 0.25$

5.26: (a)

	Cable TV	***not Cable TV***	**Total**
Internet service	250	170	420
not Internet service	550	30	580
Total	800	200	1000

(b) (i) $P(\text{customer subscribes to both cable TV and Internet service}) = \dfrac{250}{1000} = 0.25$

(ii)

$P(\text{customer subscribes to exactly one of cable TV or Internet service}) = \dfrac{550+170}{1000} = 0.72$

5.27: (a) (i) Mutually exclusive events have no outcomes in common, so $P(E \cap F) = 0$.

(ii) For mutually exclusive events, $P(E \cup F) = P(E) + P(F) = 0.41 + 0.23 = 0.64$

.

(b) No, events A and B are not mutually exclusive. For mutually exclusive events, $P(A \cup B) = P(A) + P(B)$. We are given that $P(A) = 0.26$ and $P(B) = 0.34$. If A and B were mutually exclusive, then $P(A \cup B) = P(A) + P(B) = 0.26 + 0.34 = 0.60$, which is different from the value of $P(A \cup B)$ that we are given.

5.28: Since the selection is done randomly, the events are independent of each other, so the multiplication rule for independent events applies. Therefore,

$P(\text{Luke chosen on day 1 and day 2}) = \dfrac{1}{40} \cdot \dfrac{1}{40} = \dfrac{1}{1600} = 0.000625$.

5.29: (a)

	Cash (M)	*Credit Card (not M)*	**Total**
Extended Warranty (E)	75	85	160
no Extended Warranty (not E)	395	445	840
Total	470	530	1000

(b) $P(M \cup E) = \frac{75+85+395}{1000} = 0.555$. In the long run, approximately 55.5% of customers will either pay with cash or purchase an extended warranty, or both.

Additional Exercises for Section 5.3

5.39: If the events *T* and *P* were mutually exclusive, we could add the probabilities to obtain $P(T \cup P)$. Since $P(T) + P(P) = 0.83 + 0.56 = 1.39$ is greater than 1 (and probabilities cannot be greater than 1), we know that *T* and *P* cannot be mutually exclusive.

5.41: (a) No, the events *O* and *N* are not independent. If they were independent, we could apply the multiplication rule for independent events to obtain $P(O \cap N) = P(O) \cdot P(N) = (0.7)(0.07) = 0.049$, which is not equal to the value of $P(O \cap N)$ given in the exercise.

(b)

	N	***not N***	**Total**
O	40	660	700
not O	30	270	300
Total	70	930	1000

(c) $P(O \cup N) = \frac{30+40+660}{1000} = 0.730$. In the long run, 73% of airline ticket purchasers will buy their ticket online or not show up for a flight, or both.

Section 5.4 Exercise Set 1

5.43: (a) The conditional probability is the 80%, because we are told that 80% *of those receiving such a citation* attended traffic school so that the citation would not appear on their permanent driving record. The key phrase is "of those receiving such a citation," which is what indicates that the 80% is a conditional probability.

(b) $P(F) = 0.20$ and $P(E \mid F) = 0.80$

5.44: (a) (i) $P(male) = \frac{77+117}{311} = 0.62$

(ii) $P(hybrid) = \frac{77+34}{311} = 0.36$

(iii) $P(hybrid \mid male) = \frac{77}{194} = 0.40$

(iv) $P(hybrid \mid female) = \frac{34}{117} = 0.29$

(v) $P(female \mid hybrid) = \frac{34}{111} = 0.31$

(b) (i) Of the Honda Civic purchasers, 62% were male.

(ii) Of the Honda Civic purchasers, 36% purchased a hybrid.

(iii) 40% of the male Honda Civic buyers purchased a hybrid.

(iv) 29% of the female Honda Civic buyers purchased a hybrid.

(v) 31% of the hybrid Honda Civic buyers were female.

(c) $P(hybrid \mid male) = \frac{77}{194} = 0.40$ and $P(male \mid hybrid) = \frac{77}{111} = 0.69$; These probabilities are not equal. The difference is that $P(hybrid \mid male)$ is the probability of selecting a hybrid vehicle when it is known that a purchaser is male, and $P(male \mid hybrid)$ is the probability of the purchaser being male in which it is known that a hybrid vehicle has been selected.

5.45: (a) $P(L) = 0.493$: The probability that the goalkeeper jumps to the left.

$P(C) = 0.063$: The probability that the goalkeeper stays in the center.

$P(R) = 0.444$: The probability that the goalkeeper jumps to the right.

$P(B \mid L) = 0.142$: The probability that, given that the goalkeeper jumps to the left, the kick was blocked.

$P(B \mid C) = 0.333$: The probability that, given that the goalkeeper stays in the center, the kick was blocked.

$P(B \mid R) = 0.126$: The probability that, given that the goalkeeper jumps to the right, the kick was blocked.

(b)

	Blocked (B)	***not Blocked (not B)***	**Total**
Goalkeeper jumped left (L)	70	423	493
Goalkeeper stayed center (C)	21	42	63
Goalkeeper jumped right (R)	56	388	444
Total	147	853	1000

(c) The probability that a penalty kick is blocked is $P(B) = \frac{147}{1000} = 0.147$.

(d) Considering the three conditional probabilities $P(B \mid L) = 0.142$, $P(B \mid C) = 0.333$, and $P(B \mid R) = 0.126$, it seems as if the best strategy is for the goalkeeper to stay in the center when defending a kick. The probability that the goal is blocked given that the goalkeeper stays in the center is the largest of these three conditional probabilities. This optimal strategy runs contrary to what is done in practice, namely, the goalkeeper stays in the center only 6.3% of the time, which is less than jumping left (49.3%) or jumping right (44.4%).

5.46: (a)

	Uses Alternative Therapies	Does Not Use Alternative Therapies	Total
HS/Less	315	7,005	7,320
College: 1-4 yrs	393	4,400	4,793
College: ≥5 yrs	120	975	1,095
Total	828	12,380	13,208

(b) (i)

$P(\text{randomly selected individual has 5 or more years of college}) = \frac{1,095}{13,208} = 0.083$

(ii) $P(\text{randomly selected individual uses alternative therapies}) = \frac{828}{13,208} = 0.063$

(iii)

$P(\text{individual uses alternative therapies} \mid \text{5 or more years of college}) = \frac{120}{1095} = 0.110$

(iv) $P(\text{individual uses alternative therapies} \mid \text{High School or less}) = \frac{315}{7320} = 0.043$

(v) $P(\text{High School or less} \mid \text{uses alternative therapies}) = \frac{315}{828} = 0.380$

(vi) $P(\text{individual uses alternative therapies} \mid \text{some college}) = \frac{513}{5888} = 0.087$

(c) For *H* and *A* to be independent, $P(H \cap A) = P(H) \cdot P(A)$. The necessary probabilities are $P(H) = \frac{7320}{13,208} = 0.554$, $P(A) = \frac{828}{13,208} = 0.063$, and $P(H \cap A) = \frac{315}{13,208} = 0.024$. Finally, $P(H) \cdot P(A) = (0.554)(0.063) = 0.0349$, which is not equal to 0.024. Therefore, the events *H* and *A* are not independent.

Additional Exercises for Section 5.4

5.51: $P(A \mid B)$ would be larger than $P(B \mid A)$ because the conditional probability $P(A \mid B)$ says *given that a randomly selected individual is a professional basketball player*, what is the probability that the individual is over six feet tall? It is more likely that a professional basketball player is over six feet tall than for an individual who is over six feet tall to be a professional basketball player (there are plenty of individuals over six feet tall who do not play professional basketball).

5.53: To find the requested probabilities, set up two tables as shown below. Combined, these two tables make a "hypothetical 1000" table for three events. Note that unconditional probabilities should use 1000 as the denominator in the probability calculation.

Telephone			
	Internet	**No Internet**	**Total**
Cable	150	60	210
No Cable	80	30	110
Total	230	90	320

No Telephone			
	Internet	**No Internet**	**Total**
Cable	100	490	590
No Cable	90	0	90
Total	190	490	680

Another approach is to use a Venn diagram for three events. Note the correspondence between the counts in the Venn diagram and the counts in the tables above. The counts in the Venn diagram sum to 1000, and the same probabilities can be calculated as when using the tables.

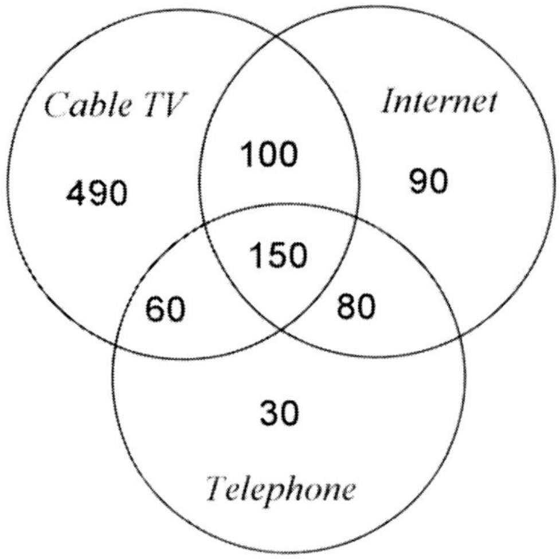

(a) $P(\text{cable TV only}) = \dfrac{490}{1000} = 0.490$; 49% of cable company customers only subscribe to cable TV.

(b) $P(\text{Internet} \mid \text{cable TV}) = \dfrac{150+100}{210+590} = \dfrac{250}{800} = 0.3125$; Of the customers who subscribe to cable TV only, 31.25% subscribe to Internet service.

(c) $P(\text{exactly two services}) = \dfrac{60+80+100}{1000} = 0.240$; 24% of cable company customers subscribe to exactly two services (cable TV and Internet, cable TV and phone, or Internet and phone).

(d) $P(\text{Internet and cable TV only}) = \dfrac{100}{1000} = 0.100$; 10% of cable company customers subscribe to Internet and cable TV only.

5.55: (a) $0.10+0.07+0.09+0.05+0.06+0.03+0.05+0.01=0.46$

(b) It more likely that he or she is in the first priority group because the proportion of students who get 11 units during the first attempt to register is higher for Priority Group 1 than it is for Priority Group 4.

(c) It is not particularly likely that a student in the third priority group would get more than 9 units during the first attempt to register because, of those in the third priority group, only 36% of students get more than 9 units. This can be found by calculating the conditional probability

$P(\text{get more than 9 units} \mid \text{third priority group}) = \dfrac{0.09}{0.25} = 0.36$.

Section 5.5 Exercise Set 1

5.57: Use of the following conditional probabilities can provide justification for the given conclusion. $P(\text{very harmful} \mid \text{current smoker}) = \dfrac{60}{96} = 0.625$,

$P(\text{very harmful} \mid \text{former smoker}) = \dfrac{78}{99} = 0.788$, and

$P(\text{very harmful} \mid \text{never smoked}) = \dfrac{86}{99} = 0.869$. The smallest of these probabilities is the probability that people who are current smokers perceive the risk of smoking as very harmful, which is consistent with the stated conclusion.

Section 5.6 Exercise Set 1

5.59: (a) $P(\text{on-time delivery in Los Angeles}) \approx \dfrac{425}{500} = 0.85$

(b) $P(\text{late delivery in Washington, D.C.}) \approx \dfrac{95}{500} = 0.19$

(c)

$$P(\text{both of two letters mailed in New York are delivered on time}) \approx \frac{415}{500} \cdot \frac{415}{500} = 0.6889$$

(d) $P(\text{on-time delivery nationwide}) \approx \dfrac{5220}{6000} = 0.87$

5.60: (a) Answers will vary. One possible way to carry out the simulation is described here. Assign each company/individual a number between 01 and 20. The table below shows one such assignment of the twenty companies or individuals, and the number of licenses requested for each company/individual, to pairs of digits.

Companies/Individuals	**Number of Licenses Requested**
01, 02, 03, 04, 05, 06	3 licenses each
07, 08, 09, 10, 11, 12, 13, 14, 15	2 licenses each
16, 17, 18, 19, 20	1 license each

Denote the individual who wants to be an independent driver, and thus have a single license, as #20. Using a random digits generator or table, read off pairs of numbers that represent the companies/individuals, and thus also represents the corresponding number of licenses requested. Ignore all pairs of numbers that are not within the range 01-20. Keep reading pairs of numbers until all 10 licenses have been issued, and determine whether or not individual #20 received a license. Repeat the simulation many times. The approximate probability that the individual's request will be granted is computed by:

$$\frac{\text{The number of times Individual \#20 was granted a license}}{\text{The total number of runs of the simulation}}.$$

After carrying out the simulation 20 times, 6 of the 20 runs granted Individual #20 a license, resulting in an approximate probability of 6/20 = 0.3. More runs of the simulation will give a more accurate approximation, and a different set of 20 runs would give a different approximate probability.

(b) This is not a fair way of distributing licenses because those companies/individuals who are requesting multiple licenses are given approximately the same chance to get 2 or 3 licenses as an individual has to get a single license. Perhaps companies who request multiple licenses should be required to submit one application per license.

Additional Exercises for Section 5.6

5.63: Results from the simulation will vary. The exact answer is 0.8468.

5.65: (a) Results from the simulation will vary. Based on the results of 30 runs of a simulation, $P(\text{more than five pairs}) \approx \frac{13}{30} = 0.43$.

(b) Results from the simulation will vary. Based on the results of 30 runs of a simulation,

$P(\text{researchers will incorrectly conclude that Treatment 2 is better}) \approx \frac{5}{30} = 0.17$.

Are You Ready to Move On? Chapter 5 Review Exercises

5.67: This probability interpretation is misleading because it demonstrates a misunderstanding of the law of averages. A better interpretation is that on average a passenger will be safe on 11,000,000 flights. Some passengers will be safe on more flights, and others on fewer flights.

5.69: (a) $P(\text{selected individual owns shares in the balanced fund}) = 0.07$

(b) $P(\text{the individual owns shares in a bond fund}) = 0.15 + 0.10 + 0.05 = 0.30$.

(c)
$P(\text{selected individual does not own shares in a stock fund}) = 1 - (0.18 + 0.25) = 0.57$

5.71: (a) $P(\text{win game 1}) = \frac{1}{50} = 0.02$ and $P(\text{win game 2}) = 2 \cdot \left(\frac{1}{10} \cdot \frac{1}{9}\right) = \frac{2}{90} = 0.022$, therefore it is advantageous to play game 2 because the probability of winning is higher.

(b) In playing either of these games 100 times, you would expect to lose money overall because in either case you can expect to lose money for each \$1 wager.

5.73: The outcomes *selected smoker who is trying to quit uses a nicotine aid* and *selected smoker who has attempted to quit begins smoking again within 2 weeks* are dependent events. An individual smoker who is trying to quit smoking is somewhat more likely to begin smoking again within two weeks if she or he does not use a nicotine aid (62%) than if she or he does use a nicotine aid (60%). The method (nicotine aid or no nicotine aid) has an impact on the success of the individual who is trying to quit smoking.

5.75: (a) $P(S)=0.987$, $P(A)=0.354$, and $P(S \cap A)=0.347$

(b)

	Specialist Opinion (***S***)	No Specialist Opinion (***not S***)	**Total**
Pathology results (***A***)	347	7	354
No Pathology results (***not A***)	640	6	646
Total	987	13	1000

(c) (i)

$$P(\text{neither a specialist opinion nor assessment of pathology results})=\frac{6}{1000}=0.006$$

(ii)

$$P(\text{specialist opinion or an assessment of pathology results})=\frac{347+640+7}{1000}=0.994$$

5.77: (a) (i) $P(E \cap F)=0$

(ii) $P(E \cup F)=0.64+0.17=0.81$

(b) A and B are not mutually exclusive events. If they were mutually exclusive, $P(A \cap B)=0$, which contradicts what is given in the exercise.

(c) A and B are not mutually exclusive events. If they were mutually exclusive, $P(A \cup B)=P(A)+P(B)=0.65+0.57=1.22$. This sum is greater than 1, which violates the rule of probability that states that probabilities are numbers between 0 and 1 (inclusive).

5.79: (a)

	Expedited Shipping (***E***)	No Expedited Shipping (***not E***)	**Total**
Gift Wrap (***G***)	31	89	120
No Gift Wrap (***not G***)	229	651	880
Total	260	740	1000

(b) $P(E \cup G) = \frac{31 + 229 + 89}{1000} = \frac{349}{1000} = 0.349$. In the long run, 34.9% of customers will request expedited shipping or gift wrap, or both.

5.81: (a) There are two conditional probabilities. One is the 39.9% who say they sometimes use Twitter to communicate with students, and the other is the 27.5% who say they sometimes use Twitter as a learning tool in the classroom. The probabilities are conditional because the description of them begins with the phrase "*Of those who use Twitter… .*"

(b) (i) $P(T) = 0.307$

(ii) $P(T^c) = 0.693$

(iii) $P(C \mid T) = 0.399$

(iv) $P(L \mid T) = 0.275$

(c)

	Uses Twitter (***T***)	Does Not Use Twitter (***not T***)	**Total**
Uses Twitter to communicate with Students (***C***)	122	0	122
Does not use Twitter to communicate with students (***not C***)	185	693	878
Total	307	693	1000

The probability that the selected study participant sometimes uses Twitter to communicate with students is $P(C) = \frac{122}{1000} = 0.122$.

(d)

	Uses Twitter (*T*)	Does Not Use Twitter (***not T***)	**Total**
Uses Twitter as a learning tool in classroom (***L***)	84	0	84
Does not use Twitter as a learning Tool in classroom (***not L***)	223	693	916
Total	307	693	1000

The probability that the selected study participant sometimes uses Twitter as a learning tool in the classroom is $P(L) = \frac{84}{1000} = 0.084$.

5.83: (a)

	Confidence High (***H***)	Confidence Low (***not H***)	**Total**
Diagnosis Correct (***C***)	98	163	261
Diagnosis Incorrect (***I***)	54	685	739
Total	152	848	1000

(b) $P(C \mid H) = \frac{98}{152} = 0.645$

(c)

	Confidence High (***H***)	Confidence Low (***not H***)	**Total**
Diagnosis Correct (***C***)	266	229	495
Diagnosis Incorrect (***I***)	127	378	505
Total	393	607	1000

$$P(C \mid H) = \frac{266}{393} = 0.677$$

This probability is higher than the probability computed in part (b). This indicates that a faculty member with high confidence in a correct diagnosis is more likely to be correct than a medical student who is also highly confident in her/his diagnosis.

Chapter 6 Random Variables and Probability Distributions

Section 6.1 Exercise Set 1

6.1: (a) discrete

(b) continuous

(c) discrete

(d) discrete

(e) continuous

6.2: The possible values for x are $x \geq 1$ (the positive integers).

Five possible outcomes, with their corresponding x values, are shown below.

Outcomes	x
S	1
LS	2
RLS	3
RRS	3
LRLRS	5

6.3: (a) 3, 4, 5, 6, 7

(b) –3, –2, –1, 1, 2, 3

(c) 0, 1, 2

(d) 0, 1

Additional Exercises for Section 6.1

6.7: (a) discrete

(b) continuous

(c) continuous

(d) discrete

6.9: (a) 2, 3, 4, 5, 6, 7, 8, 9, 10, 11, 12

(b) –5, –4, –3, –2, –1, 0, 1, 2, 3, 4, 5

(c) 1, 2, 3, 4, 5, 6

Section 6.2 Exercise Set 1

6.10: (a) $p(4) = 1-(0.65+0.20+0.10+0.04) = 0.01$

(b) $p(1) = 0.20$ is interpreted as the probability of randomly selecting a carton of one dozen eggs and finding exactly 1 broken egg is 0.20.

(c) $P(y \le 2) = p(0) + p(1) + p(2) = 0.65 + 0.20 + 0.10 = 0.95$; The probability that a randomly selected carton of eggs contains 0, 1, or 2 broken eggs is 0.95.

(d) $P(y < 2) = p(0) + p(1) = 0.65 + 0.20 = 0.85$; This probability is smaller than the probability in part (c) because we only want fewer than 2 broken eggs, not two or fewer as was the case in part (c). Therefore, we are only considering 0 or 1 broken egg.

(e) Exactly 10 unbroken eggs is equivalent to exactly 2 broken eggs. $p(2) = 0.10$.

(f) At least 10 unbroken eggs is equivalent to 10, 11, or 12 unbroken eggs, or 2, 1, or 0 broken eggs. From Part (c), $P(y \le 2) = 0.95$.

6.11: (a) For 1000 graduates, we would expect to see approximately 450 graduates who contributed nothing (either through declining to make a donation or were not reached by phone), 300 graduates to contribute $10, 200 graduates to contribute $25, and 50 graduates to contribute $50.

(b) The most common value of *x* in this population is $0.

(c) $P(x \ge 25) = p(25) + p(50) = 0.20 + 0.05 = 0.25$

(d) $P(x > 0) = p(10) + p(25) + p(50) = 0.30 + 0.20 + 0.05 = 0.55$

6.12: (a) (1, 2), (1, 3), (1, 4), (2, 3), (2, 4), (3, 4)

(b) Because the bottles are randomly selected, each of the outcomes enumerated in Part (a) are equally likely. Therefore, each has probability 1/6.

(c) The x value for each possible outcome is shown in the table below.

Outcome	x
(1, 2)	0
(1, 3)	1
(1, 4)	1
(2, 3)	1
(2, 4)	1
(3, 4)	2

The probability distribution of x is:

x	0	1	2
$p(x)$	1/6	4/6	1/6

6.13: (a) The table below shows all 16 possible outcomes of homeowners with earthquake insurance (S), and those without earthquake insurance (F), for four homeowners. Also shown are the computed probabilities, and the value of x, the number among the four who have earthquake insurance.

Outcome	Probability	x
SSSS	(0.2)(0.2)(0.2)(0.2) = 0.0016	4
SSSF	(0.2)(0.2)(0.2)(0.8) = 0.0064	3
SSFS	(0.2)(0.2)(0.8)(0.2) = 0.0064	3
SFSS	(0.2)(0.8)(0.2)(0.2) = 0.0064	3
FSSS	(0.8)(0.2)(0.2)(0.2) = 0.0064	3
SSFF	(0.2)(0.2)(0.8)(0.8) = 0.0256	2
SFSF	(0.2)(0.8)(0.2)(0.8) = 0.0256	2
SFFS	(0.2)(0.8)(0.8)(0.2) = 0.0256	2
FFSS	(0.8)(0.8)(0.2)(0.2) = 0.0256	2
FSSF	(0.8)(0.2)(0.2)(0.8) = 0.0256	2
FSFS	(0.8)(0.2)(0.8)(0.2) = 0.0256	2
FFFS	(0.8)(0.8)(0.8)(0.2) = 0.1024	1
FFSF	(0.8)(0.8)(0.2)(0.8) = 0.1024	1
FSFF	(0.8)(0.2)(0.8)(0.8) = 0.1024	1
SFFF	(0.2)(0.8)(0.8)(0.8) = 0.1024	1
FFFF	(0.8)(0.8)(0.8)(0.8) = 0.4096	0

The probability distribution of x is shown in the table below. The values in the probability distribution are found by adding the probabilities for the values of x from the table above.

x	0	1	2	3	4
$p(x)$	0.4096	0.4096	0.1536	0.0256	0.0016

(b) The most likely outcomes are 0 and 1.

(c) $P(x \geq 2) = p(2) + p(3) + p(4) = 0.1536 + 0.0256 + 0.0016 = 0.1808$

Additional Exercises for Section 6.2

6.19: (a) Because $p(y) = ky$, $y = 1,\ldots,5$, and $\sum ky = 1$, we know that

$k(1) + k(2) + k(3) + k(4) + k(5) = 1$, or $15k = 1$, so $k = \frac{1}{15}$.

(b) The probability that at most three forms are required is

$P(y \leq 3) = p(1) + p(2) + p(3) = \frac{1}{15}(1) + \frac{1}{15}(2) + \frac{1}{15}(3) = \frac{1}{15}(6) = 0.4$

(c) The probability that between two and four forms (inclusive) are required is

$P(2 \leq y \leq 4) = p(2) + p(3) + p(4) = \frac{1}{15}(2) + \frac{1}{15}(3) + \frac{1}{15}(4) = 0.6$

Section 6.3 Exercise Set 1

6.21: (a)

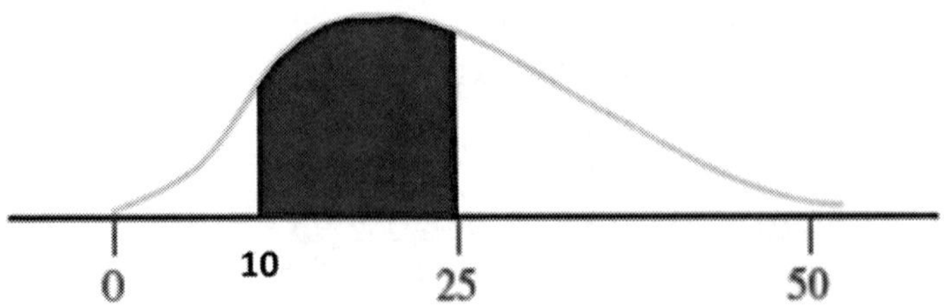

(b)

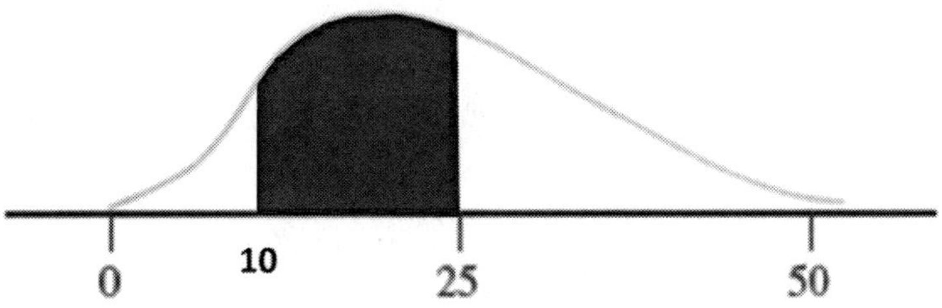

(c)

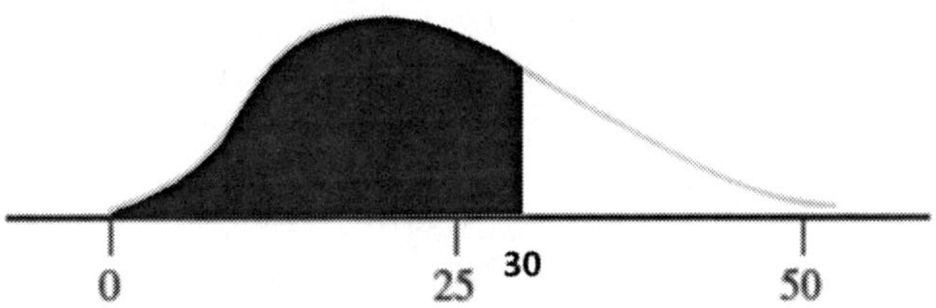

(d)

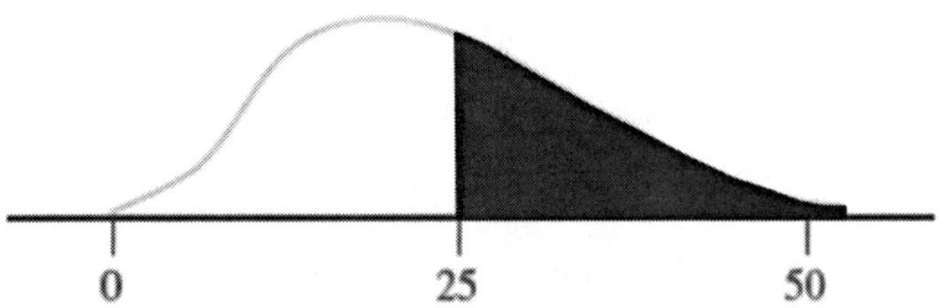

6.22 (a) The probability that the time to wait for service is between 4 and 7 minutes, or $P(4 \le x \le 7)$.

(b) The probability that a randomly selected individual waits between 4 and 7 minutes for service at a bank is 0.26.

6.23: (a) The density curve creates a triangular region bounded by the line, the y-axis, and the x-axis. Since the area of the triangular region must equal one, and the formula for the area of a triangle is $A = \frac{1}{2} b \cdot h$, with the given base of $b = 1$, $\frac{1}{2} \cdot 1 \cdot h = 1$, so $h = 2$.

(b) $P(x > 0.5) = \frac{1}{2}(0.5)(1) = 0.25$, using the formula for the area of the triangle formed for $x > 0.5$.

(c) $P(x \le 0.25) = 1 - P(x > 0.25) = 1 - \frac{1}{2}(0.75)(1.5) = 0.4375$.

6.24: (a) $A_{total} = \frac{1}{2}(0.40)(5) = 1$

(b) By dividing the region into triangles, and using the formula $A = \frac{1}{2} b \cdot h$,

$P(x < 0.20) = 0.5$; $P(x < 0.1) = \frac{1}{2}(0.1)(2.5) = 0.125$;

$P(x > 0.3) = \frac{1}{2}(0.1)(2.5) = 0.125$

(c) $P(0.10 < x < 0.20) = 1 - \left(P(x < 0.10) + P(x > 0.20)\right) = 1 - (0.125 + 0.5) = 0.375$

Additional Exercises for Section 6.3

6.29: The probability $P(x < 1)$ is the smallest. $P(x > 3)$ and $P(2 < x < 3)$ are equal to each other, and larger than the other two probabilities.

6.31: $P(2 < x < 3) = P(2 \le x \le 3) < P(x < 2) < P(x > 7)$. The reason for this ordering of the probabilities is that the smallest two probabilities are both equal to 1/10, the third probability is equal to 2/10, and the fourth probability is equal to 3/10. These values can be determined by shading the appropriate regions beneath the density curve and comparing the areas.

Section 6.4 Exercise Set 1

6.32: (a) $\mu_y = \sum y \cdot p(y) = 0(0.65) + 1(0.20) + 2(0.10) + 3(0.04) + 4(0.01) = 0.56$; This is the mean value of the number of broken eggs in the population of egg cartons.

(b) The only value of y that is less than 0.56 is 0, therefore $P(y < 0.56) = P(y = 0) = 0.65$. This is not particularly surprising because, in the long run, 65% of egg cartons contain no broken eggs.

(c) This computation of the mean is incorrect because it does not take into account the probabilities with which the number of broken eggs need to be weighted. As written, the computation indicates that the number of broken eggs (0, 1, 2, 3, or 4) are all equally likely.

6.33: (a) $\mu_x = \sum x \cdot p(x) = 13.5(0.2) + 15.9(0.5) + 19.1(0.3) = 16.38$ and

$$\sigma_x^2 = \sum (x - \mu_x)^2 \cdot p(x) = (13.5 - 16.38)^2(0.2) + (15.9 - 16.38)^2(0.5) + (19.1 - 16.38)^2(0.3)$$
$$= (-2.88)^2(0.2) + (-0.48)^2(0.5) + (2.72)^2(0.3) = 3.9936$$

Therefore, $\sigma_x = \sqrt{\sigma_x^2} = \sqrt{3.9936} = 1.9984$

(b) The mean, $\mu_x = 16.38$ cubic feet represents the long run average storage space of freezers sold by this particular appliance dealer. The standard deviation, $\sigma_x = 1.9984$ cubic feet, represents a typical deviation in how much the storage space in freezers purchased deviates from the mean.

6.34: Answers may vary. Two possible probability distributions are shown below.

Probability Distribution 1 ($\mu_x = 3$ and $\sigma_x = 1.265$):

x	1	2	3	4	5
p(x)	0.15	0.20	0.30	0.20	0.15

Probability Distribution 2 ($\mu_x = 3$ and $\sigma_x = 1.643$):

x	1	2	3	4	5
p(x)	0.30	0.15	0.10	0.15	0.30

Additional Exercises for Section 6.4

6.39: (a) The mean is $\mu_x = \sum x \cdot p(x) = 1(0.2) + 2(0.4) + 3(0.3) + 4(0.1) = 2.3$ and represents the long run average number of lots ordered per customer.

(b) The variance is

$$\sigma_x^2 = \sum (x-\mu_x)^2 \cdot p(x) = (1-2.3)^2 \cdot (0.2) + (2-2.3)^2 \cdot (0.4)$$
$$+ (3-2.3)^2 \cdot (0.3) + (4-2.3)^2 \cdot (0.1) = 0.81$$

square lots, and the standard deviation is $\sigma_x = \sqrt{\sigma_x^2} = \sqrt{0.81} = 0.9$ lots. The variance measures how much, on average, the squared number of lots ordered per customer deviates from the mean. The standard deviation measures how much, on average, the number of lots ordered per customer deviates from the mean.

Section 6.5 Exercise Set 1

6.41: (a) The area under the z curve to the left of 1.75 is $P(z \le 1.75) = 0.9599$

(b) The area under the z curve to the left of -0.68 is $P(z \le -0.68) = 0.2483$

(c) The area under the z curve to the right of 1.20 is
$P(z \ge 1.20) = 1 - P(z \le 1.20) = 1 - 0.8849 = 0.1151$

(d) The area under the z curve to the right of -2.82 is
$P(z \ge -2.82) = 1 - P(z \le -2.82) = 1 - 0.0024 = 0.9976$

(e) The area under the z curve between -2.22 and 0.53 is
$P(-2.22 \le z \le 0.53) = P(z \le 0.53) - P(z \le -2.22) = 0.7019 - 0.0132 = 0.6887$

(f) The area under the z curve between -1 and 1 is
$P(-1 \le z \le 1) = P(z \le 1) - P(z \le -1) = 0.8413 - 0.1587 = 0.6826$

(g) The area under the z curve between -4 and 4 is
$P(-4 \le z \le 4) = P(z \le 4) - P(z \le -4) = 0.999968 - 0.000032 = 0.999936$ (using technology)

6.42: (a) $P(z < 2.36) = 0.9909$

(b) $P(z \le 2.36) = 0.9909$

(c) $P(z < -1.23) = 0.1093$

(d) $P(1.14 < z < 3.35) = P(z < 3.35) - P(z < 1.14) = 0.9996 - 0.8729 = 0.1267$

(e) $P(-0.77 \le z \le -0.55) = P(z \le -0.55) - P(z \le -0.77) = 0.2912 - 0.2206 = 0.0706$

(f) $P(z > 2) = 1 - P(z \le 2) = 1 - 0.9772 = 0.0228$

(g) $P(z \ge -3.38) = 1 - P(z < -3.38) = 1 - 0.0004 = 0.9996$

(h) $P(z < 4.98) \approx 1$

6.43: (a) $P(x < 5) = P\left(z < \dfrac{5-5}{0.2}\right) = P(z < 0) = 0.5$

(b) $P(x < 5.4) = P\left(z < \dfrac{5.4-5}{0.2}\right) = P(z < 2) = 0.9772$

(c) $P(x \le 5.4) = P\left(z \le \dfrac{5.4-5}{0.2}\right) = P(z \le 2) = 0.9772$

(d)

$$P(4.6 < x < 5.2) = P\left(\frac{4.6-5}{0.2} < z < \frac{5.2-5}{0.2}\right) = P(-2 < z < 1) = P(z < 1) - P(z < -2)$$
$$= 0.8413 - 0.0228 = 0.8185$$

(e)

$$P(x > 4.5) = P\left(z > \frac{4.5-5}{0.2}\right) = P(z > -2.5) = 1 - P(z \le -2.5) = 1 - 0.0062 = 0.9938$$

(f) $P(x > 4.0) = 1 - P(x \le 4.0) = 1 - P\left(z \le \dfrac{4.0-5.0}{0.2}\right) = 1 - P(z \le -5.0) \approx 1 - 0 = 1$

6.44: Let s be the typist's speed in words per minute.

(a) At most 60 wpm: $P(s \le 60) = P\left(z \le \dfrac{60-60}{15}\right) = P(z \le 0) = 0.5$; Less than 60 wpm: $P(s < 60) = P\left(z < \dfrac{60-60}{15}\right) = P(z < 0) = 0.5$

(b)
$$P(45 \le s \le 90) = P(s \le 90) - P(s \le 45) = P\left(z \le \frac{90-60}{15}\right) - P\left(z \le \frac{45-60}{15}\right)$$
$$= P(z \le 2) - P(z \le -1) = 0.9772 - 0.1587 = 0.8185$$

(c) $P(s > 105) = P\left(z > \frac{105-60}{15}\right) = P(z > 3) = 0.0013$; It would be surprising to find a typist in this population whose speed exceeded 105 wpm because the probability of finding such a typist is only 0.13%.

(d) The probability of finding one typist with a typing speed that exceeds 75 wpm is $P(s > 75) = P\left(z > \frac{75-60}{15}\right) = P(z > 1) = 0.1587$. The probability that both typists have typing speeds that exceed 75 wpm is (0.1587)(0.1587) = 0.0252 (by the multiplication rule for independent events).

(e) We need the value of z^* such that $P(z \le z^*) = 0.20$. Using technology, $z^* = -0.8416$. Therefore, $-0.8416 = \frac{s-60}{15} \Rightarrow s = (-0.8416)(15) + 60 = 47.376$. Any typist with a typing speed of 47.376 wpm or less is in the slowest 20% of typists.

6.45: The proportion of corks that are not defective is found by $P(2.9 < x < 3.1) = P\left(\frac{2.9-3}{0.1} < z < \frac{3.1-3}{0.1}\right) = P(-1 < z < 1) = 0.6827$. Therefore, the proportion of corks produced by this machine that are defective is $1 - 0.6827 = 0.3173$.

6.46: The proportion of corks that are not defective is $P(2.9 < y < 3.1) = P\left(\frac{2.9-3.05}{0.01} < z < \frac{3.1-3.05}{0.01}\right) = P(-15 < z < 5) \approx 1$. Therefore, the proportion of corks produced by this machine that are defective is approximately $1 - 1 = 0$. The second machine produces fewer defective corks.

Additional Exercises for Section 6.5

6.53: It is not reasonable to think that traffic flow is approximately normal because traffic flow cannot be negative, and 0 traffic flow is 1.58 standard deviations below the mean $\left(z = \frac{0-0.41}{0.26} = -1.58\right)$, which indicates that approximately 5.7% of traffic flow values would be negative if the traffic times actually followed a normal distribution.

6.55: Since these values are times, they must all be positive. In the normal distribution with mean 9.9 and standard deviation 6.2, approximately 5.5% of processing times would be less than or equal to zero.

6.57: Let w represent the weight of mozzarella cheese on a medium pizza.

(a) $P(0.525 < w < 0.550) = P\left(\frac{0.525-0.5}{0.025} < z < \frac{0.550-0.5}{0.025}\right) = P(1 < z < 2) = 0.136$

(b) $P(z > 2) = 1 - P(z \le 2) = 1 - 0.9772 = 0.0228$

(c) Let w_1, w_2, and w_3 represent the weight of the cheese on the three pizzas. The probability that three randomly selected medium pizzas all have at least 0.475 lb of cheese is $P\left((w_1 \ge 0.475) \cap (w_2 \ge 0.475) \cap (w_3 \ge 0.475)\right)$. Since the pizzas were selected randomly, they are independent of each other, so, by applying the multiplication rule for independent events, the probability is computed as $P(w_1 \ge 0.475) \cdot P(w_2 \ge 0.475) \cdot P(w_3 \ge 0.475)$. These probabilities are all equal to each other, so

$$P(w \ge 0.475) = P\left(z \ge \frac{0.475-0.5}{0.025}\right) = P(z \ge -1) = 1 - P(z < -1) = 1 - 0.1587 = 0.8413$$

, and $P(w_1 \ge 0.475) \cdot P(w_2 \ge 0.475) \cdot P(w_3 \ge 0.475) = (0.8413)^3 = 0.5955$.

6.59: (a)

$$P(x > 45) = P\left(z > \frac{45-60}{10}\right) = P(z > -1.5) = 1 - P(z \le -1.5) = 1 - 0.0668 = 0.9332$$

(b) We need the value of z^* such that $P(z > z^*) = 0.10$. This is equivalent to $P(z \le z^*) = 0.90$, the 90th percentile. Using technology, we find that $P(z \le 1.282) = 0.90$. Therefore, $1.282 = \frac{x-60}{10} \Rightarrow x = 10(1.282) + 60 = 72.82$ minutes.

6.61: Let x represent the amount of vitamin E in each capsule.

Therefore, $P(x < 4.9) = P\left(z < \frac{4.9-5}{0.05}\right) = P(z < -2) = 0.0228$;

$$P(x \ge 5.2) = P\left(z \ge \frac{5.2-5}{0.05}\right) = P(z \ge 4) \approx 0$$

6.63: We want to find z^* such that $P(z > z^*) = 0.15$. This is equivalent to $P(z \le z^*) = 0.85$, the 85th percentile. Using technology, we find that

$P(z \le 1.036) = 0.85$. Therefore, $1.036 = \frac{x - 78}{7} \Rightarrow x = 7(1.036) + 78 = 85.252$.

Since your score of 89 is higher than 85.252, you received an A.

6.65: We want to find z^* such that $P(z \le z^*) = 0.20$. Using a table of standard normal probabilities or technology, we find that $P(z \le -0.842) = 0.20$. Therefore,

$-0.842 = \frac{x - 700}{50} \Rightarrow x = 50(-0.842) + 700 = 657.9$ hours. The bulbs should be replaced every 657.9 hours so that no more than 20% of the bulbs will have already burned out.

Section 6.6 Exercise Set 1

6.67: (a) The plot (shown below) does not look linear. The nonlinearity of the normal probability plot supports the author's assertion that the distribution of fussing times is not normal.

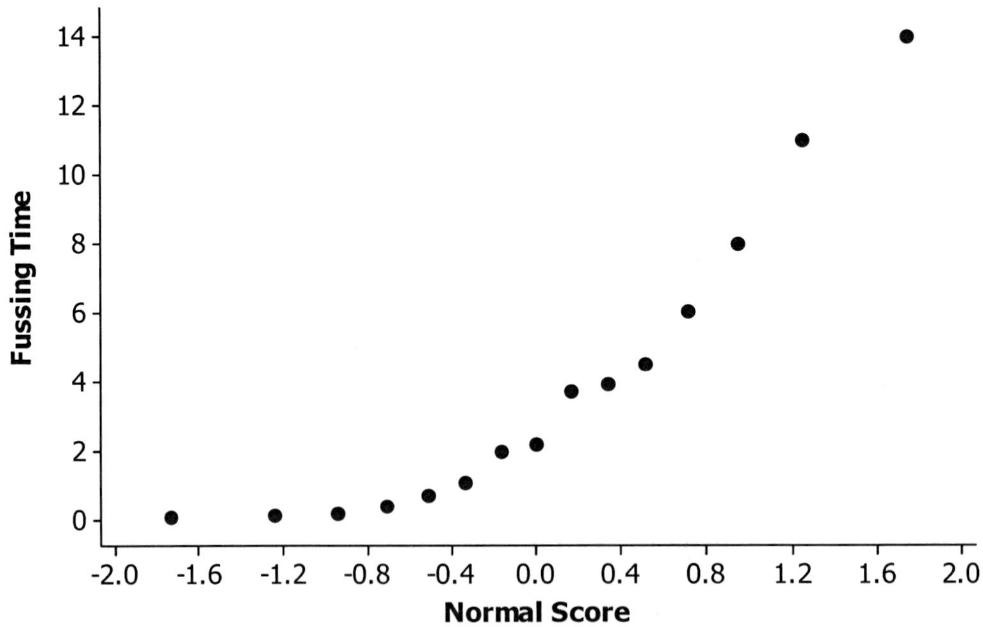

(b) The computed correlation coefficient is $r = 0.920$. The *critical r* from Table 6.2 is 0.911. Since the computed *r* is greater than the *critical r*, it is reasonable to think that the population distribution is normal.

6.68: (a)

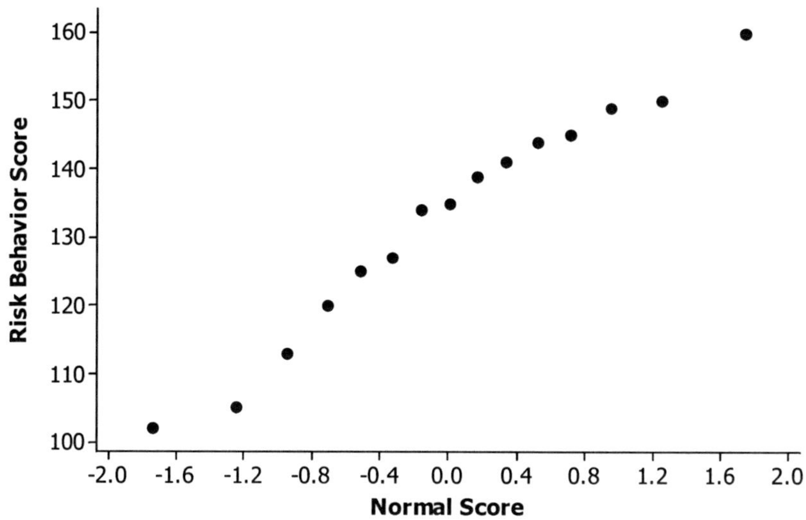

(b)

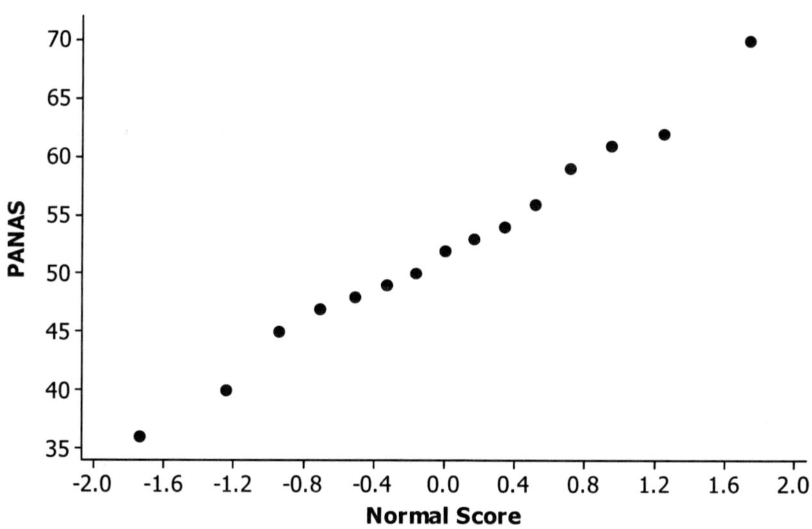

(c) Because both normal probability plots are approximately linear, it seems reasonable that both risk behavior scores and PANAS scores are approximately normally distributed. In addition, the correlation coefficients for both of the plots ($r_{risk} = 0.985$ and $r_{PANAS} = 0.993$) are both above the *critical r* cutoff from Table 6.2 (*critical r* = 0.911), which provides further justification for the claim of normality.

Additional Exercises for Section 6.6

6.71: (a) Yes, the normal probability plot appears linear.

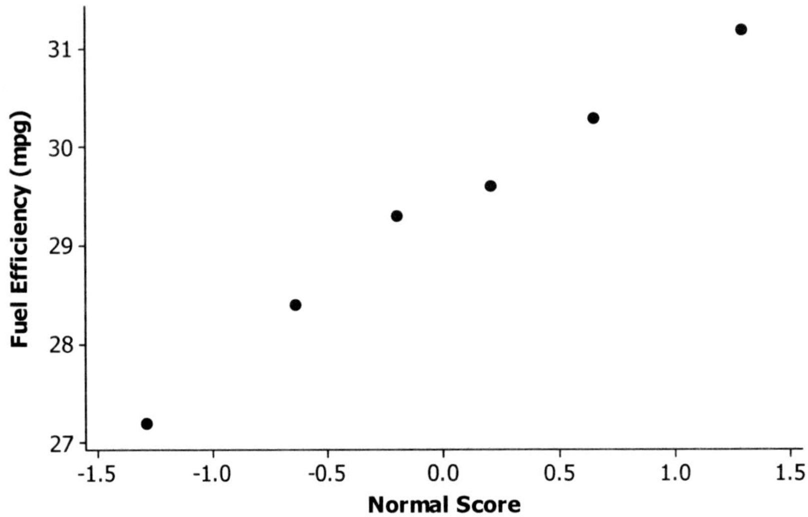

(b) $r = 0.994$; From Table 6.2, we find that the *critical r* for $n = 6$ lies between 0.832 (*critical r* for $n = 5$) and 0.880 (*critical r* for $n = 10$). The computed correlation coefficient is larger than the *critical r* for $n = 6$, so it is reasonable to think that the fuel efficiency distribution is approximately normal.

6.73:

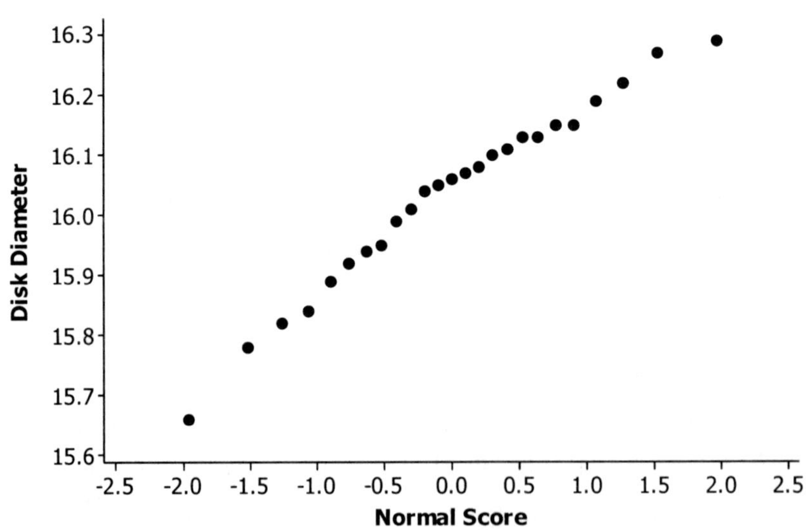

Based on the linearity of the normal probability plot and the fact that the correlation coefficient for the (normal score, disk diameter) points ($r = 0.987$) exceeds the *critical r* of 0.941 (from Table 6.2), it is reasonable to think that disk diameter is normally distributed.

Section 6.7 Exercise Set 1

6.74: (a) x can take on the values 0, 1, 2, 3, 4, or 5.

(b) The probability for each value of x is computed using the formula $p(x) = \frac{n!}{x!(n-x)!}p^x(1-p)^{n-x} = \frac{5!}{x!(5-x)!}\left(\frac{1}{4}\right)^x\left(\frac{3}{4}\right)^{5-x}$. For example, two of the six probabilities are calculated as follows:

$p(0) = \frac{5!}{0!(5-0)!}\left(\frac{1}{4}\right)^0\left(\frac{3}{4}\right)^{5-0} = (1)(1)(0.75)^5 = 0.2373$ and

$p(2) = \frac{5!}{2!(5-2)!}\left(\frac{1}{4}\right)^2\left(\frac{3}{4}\right)^{5-2} = (10)(0.25)^2(0.75)^3 = 0.2637$. The remaining probabilities are calculated similarly and the results are shown in the table below.

x	0	1	2	3	4	5
$p(x)$	0.2373	0.3955	0.2637	0.0879	0.0146	0.0010

(c)

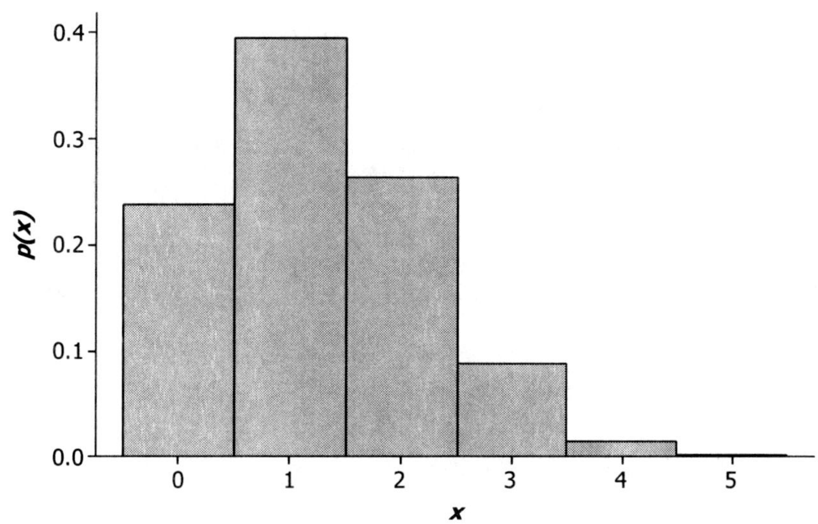

6.75: (a) $P(x=2)=p(2)=\dfrac{4!}{2!(4-2)!}(0.9)^2(0.1)^2=0.0486$; The probability that exactly two of the four randomly selected households have cable TV is 0.0486.

(b) $P(x=4)=p(4)=\dfrac{4!}{4!(4-4)!}(0.9)^4(0.1)^0=0.6561$

(c) $P(x\le 3)=1-P(x=4)=1-0.6561=0.3439$

6.76: (a) $P(x=2)=p(2)=\dfrac{5!}{2!(5-2)!}(0.25)^2(0.75)^3=0.2637$

(b) $$P(x\le 1)=p(0)+p(1)=\frac{5!}{0!(5-0)!}(0.25)^0(0.75)^5+\frac{5!}{1!(5-1)!}(0.25)^1(0.75)^4$$
$$=0.2373+0.3955=0.6328$$

(c) $P(2\le x)=1-P(x\le 1)=1-0.6328=0.3672$

(d) $P(x\ne 2)=1-P(x=2)=1-0.2637=0.7363$

6.77: (a) Binomial distribution with $n = 20$ and $p = 0.05$. Accept if shipped lot contains 0 or 1 defective part. Therefore,

$$p(0)+p(1)=\frac{20!}{0!(20-0)!}(0.05)^0(0.95)^{20}+\frac{20!}{1!(20-1)!}(0.05)^1(0.95)^{19}$$
$$=0.3585+0.3774=0.7359$$

(b) Binomial distribution with $n = 20$ and $p = 0.10$. Accept if shipped lot contains 0 or 1 defective part. Therefore,

$$p(0)+p(1)=\frac{20!}{0!(20-0)!}(0.10)^0(0.90)^{20}+\frac{20!}{1!(20-1)!}(0.10)^1(0.90)^{19}$$
$$=0.1216+0.2702=0.3918$$

(c) Binomial distribution with $n = 20$ and $p = 0.20$. Accept if shipped lot contains 0 or 1 defective part. Therefore,

$$p(0)+p(1)=\frac{20!}{0!(20-0)!}(0.20)^0(0.80)^{20}+\frac{20!}{1!(20-1)!}(0.20)^1(0.80)^{19}$$
$$=0.0115+0.0576=0.0691$$

6.78: (a) The probability distribution of x is not binomial because the question asks about songs being played *until* a song by this particular artist is played. There is not a fixed number of trials, which is required for the binomial distribution. This setting is geometric.

(b) (i) $p(4) = (1-0.08)^3(0.08) = 0.0623$

(ii) $P(x \le 4) = p(1) + p(2) + p(3) + p(4)$
$= (0.92)^0(0.08) + (0.92)^1(0.08) + (0.92)^2(0.08) + (0.92)^3(0.08)$
$= 0.2836$

(iii) $P(x > 4) = 1 - P(x \le 4) = 1 - 0.2836 = 0.7164$

(iv) $P(x \ge 4) = p(4) + P(x > 4) = 0.0623 + 0.7164 = 0.7787$

(c) The differences between the four probabilities are shown in bold font.

(i) The probability that it takes **exactly four songs** until the first song by the particular artist is played is 0.0623.

(ii) The probability that it takes **at most four songs** until the first song by the particular artist is played is 0.2836.

(iii) The probability that it takes **more than four songs** until the first song by the particular artist is played is 0.7164.

(iv) The probability that it takes **at least four songs** until the first song by the particular artist is played is 0.7787.

6.79: (a) The probability distribution of x is geometric, with success probability $p = 0.15$.

(b) $P(x = 3) = (1-0.15)^2(0.15) = 0.1084$

(c) $P(x < 4) = p(1) + p(2) + p(3) = 0.15 + (0.85)(0.15) + (0.85)^2(0.15) = 0.3859$

(d) $P(x > 3) = 1 - P(x \le 3)$
$= 1 - [p(1) + p(2) + p(3)]$
$= 1 - [0.15 + (0.85)(0.15) + (0.85)^2(0.15)] = 0.6141$

Additional Exercises for Section 6.7

6.87: The binomial probability distribution with $n = 5$ and $p = 0.5$ is

x	0	1	2	3	4	5
$p(x)$	0.03125	0.15625	0.31250	0.31250	0.15625	0.03125

6.89: This scenario is sampling without replacement. Since more than 5% of the population is being sampled (the percentage of the population being sampled is 2,000/10,000 = 20%), the binomial distribution will not give a good approximation of the probability distribution of the number of invalid signatures.

6.91: (a) This is a binomial setting, with $n = 25$ and $p = 0.5$. The probability that the coin is judged biased is $P(x \le 7) + P(x \ge 18) = 0.0216 + 0.0216 = 0.0432$.

(b) This is a binomial setting, with $n = 25$ and $p = 0.9$. The probability that the coin is judged biased is $P(x \le 7) + P(x \ge 18) = 0.000 + 0.9977 = 0.9977$. Therefore, the probability that the coin is judged fair is $1 - 0.9977 = 0.0023$.

Changing the scenario to $P(\text{H}) = 0.1$, use the binomial setting with $n = 25$ and $p = 0.1$. Therefore, the probability that the coin is judged biased is $P(x \le 7) + P(x \ge 18) = 0.9977 + 0.0000 = 0.9977$. Therefore, the probability that the coin is judged fair is $1 - 0.9977 = 0.0023$.

(c) With $P(\text{H}) = 0.6$, use the binomial setting with $n = 25$ and $p = 0.6$. Therefore, the probability that the coin is biased is $P(x \le 7) + P(x \ge 18) = 0.0012 + 0.1536 = 0.1548$. Therefore, the probability that the coin is judged fair is $1 - 0.1548 = 0.8452$. With $P(\text{H}) = 0.4$, use the binomial setting with $n = 25$ and $p = 0.4$. Therefore, the probability that the coin is judged biased is $P(x \le 7) + P(x \ge 18) = 0.1536 + 0.0012 = 0.1548$. Therefore, the probability that the coin is judged fair is $1 - 0.1548 = 0.8452$. The probabilities are large compared to the probabilities in Part (b) because the probabilities of getting a head in Part (c) are closer to 0.5 than those in Part (b). Probabilities closer to 0.5 will make it more likely that the coin will be judged fair.

(d) Changing the definition of being judged fair to $7 \le x \le 18$ increases the likelihood that the coin will be judged fair, and decreases the probability that the coin is judged to not be fair. Although the new rule is more likely to judge a fair coin as fair, it is also more likely to judge a biased coin as fair.

Section 6.8 Exercise Set 1

6.93: (a) $P(x=100) \approx P\left(\frac{99.5-100}{15} \le z \le \frac{100.5-100}{15}\right) = P(-0.033 \le z \le 0.033) = 0.026$

(b) $P(x \le 110) \approx P\left(z \le \frac{110.5-100}{15}\right) = P(z \le 0.7) = 0.7580$

(c) $P(x < 110) = P(x \le 109) \approx P\left(z \le \frac{109.5-100}{15}\right) = P(z \le 0.633) = 0.7366$

(d) $P(75 \le x \le 125) \approx P\left(\frac{74.5-100}{15} \le z \le \frac{125.5-100}{15}\right) = P(-1.7 \le z \le 1.7) = 0.9109$

6.94: To use the normal approximation in this scenario, first verify that $np \ge 10$ and $n(1-p) \ge 10$, where n = 100 and p = 0.7. $(100)(0.7) = 70$ and $100(1-0.7) = 30$, both values are greater than 10. Second, find the mean and standard deviation of x, where x represents the number of mountain bikes out of 100 randomly selected bikes sold. The mean is $\mu_x = np = (100)(0.7) = 70$ and the standard deviation is $\sigma_x = \sqrt{np(1-p)} = \sqrt{100 \cdot 0.7 \cdot (1-0.7)} = \sqrt{21} = 4.5826$.

(a) $P(x \le 75) \approx P\left(z \le \frac{75.5-70}{4.5826}\right) = P(z \le 1.200) = 0.8849$

(b) $P(60 \le x \le 75) \approx P\left(\frac{59.5-70}{4.5826} \le z \le \frac{75.5-70}{4.5826}\right) = P(-2.291 \le z \le 1.200) = 0.8739$

(c)

$P(x > 80) = P(x \ge 81) \approx P\left(z \ge \frac{80.5-70}{4.5826}\right) = P(z \ge 2.291) = 1 - P(z \le 2.291) = 0.0110$

(d) At most 30 are not mountain bikes is the same as at least 70 are mountain bikes. Therefore, $P(x \ge 70) = P\left(z \ge \frac{69.5-70}{4.5826}\right) = P(z \ge -0.109) = 0.5434$

6.95: Check $np \ge 10$ and $n(1-p) \ge 10$ to verify that using the normal approximation is appropriate. Both $np = (225)(0.65) = 146.25$ and $n(1-p) = (225)(1-0.65) = 78.75$ are greater than 10. Let x represent the number of registered voters out of 225 in a certain area who favor a 7-day waiting period before purchasing a hand gun.

Compute the mean and standard deviation to use the normal approximation. The mean is $\mu_x = np = (225)(0.65) = 146.25$ and the standard deviation is
$\sigma_x = \sqrt{np(1-p)} = \sqrt{225 \cdot 0.65 \cdot (1-0.65)} = 7.1545$

(a) $P(x \geq 150) \approx P\left(z \geq \frac{149.5 - 146.25}{7.1545}\right) = P(z \geq 0.4543) = 0.3248$

(b) $P(x > 150) = P(x \geq 151) \approx P\left(z \geq \frac{150.5 - 146.25}{7.1545}\right) = P(z \geq 0.5940) = 0.2763$

(c) $P(x < 125) = P(x \leq 124) \approx P\left(z \leq \frac{124.5 - 146.25}{7.1545}\right) = P(z \leq -3.040) = 0.0012$

Additional Exercises for Section 6.8

6.99: (a) To verify the appropriate use of the normal approximation, both np and $n(1-p)$ must be at least 10. Both $np = (60)(0.7) = 42$ and $n(1-p) = 60(1-0.7) = 18$ are at least 10.

(b) Compute mean and standard deviation. $\mu_x = np = (60)(0.7) = 42$ and
$\sigma_x = \sqrt{np(1-p)} = \sqrt{60 \cdot 0.7 \cdot (1-0.7)} = 3.5497$

(i)

$$P(x = 42) \approx P\left(\frac{41.5 - 42}{3.5497} \leq z \leq \frac{42.5 - 42}{3.5497}\right) = P(-0.1409 \leq z \leq 0.1409) = 0.1121$$

(ii) $P(x < 42) = P(x \leq 41) \approx P\left(z \leq \frac{41.5 - 42}{3.5497}\right) = P(z \leq -0.1409) = 0.4440$

(iii) $P(x \leq 42) \approx P\left(z \leq \frac{42.5 - 42}{3.5497}\right) = P(z \leq 0.1409) = 0.5560$

(c) The probability in (i) represents the probability of exactly 42 correct responses on the MENT test for a patient who is trying to fake the test. The probability in (ii) is the probability of less than 42 correct responses on the MENT test for a patient who is trying to fake the test. The probability in (iii) represents the probability of

42 or fewer correct responses on the MENT test for a patient who is trying to fake the test. These are different numbers of correct questions.

(d) The normal approximation is not appropriate here because $n(1-p) = 60(1-0.96) = 2.4$, which is less than 10. As such, the binomial distribution must be used.

(e) The small probability in part (d) indicates that it is extremely unlikely for someone who is not faking the test to correctly answer 42 or fewer questions, when compared with the probability that a person who is faking the test to correctly answer 42 or fewer questions (0.556). If someone does score 42 or fewer points on the MENT exam, there is strong evidence (supported by the extremely small probability) that the person must be faking.

Are You Ready to Move On? Chapter 6 Review Exercises

6.101: The depth, x, can take on any value between 0 and 100 ($0 \le x \le 100$), inclusive. x is continuous.

6.103: (a) The plane can accommodate 100 passengers, so if 100 or fewer passengers show up for the flight, everyone can be accommodated. Therefore, $P(x \le 100) = 0.05 + 0.10 + 0.12 + 0.14 + 0.24 + 0.17 = 0.82$.

(b) This is the complement of all passengers being accommodated, or $P(x > 100) = 1 - P(x \le 100) = 1 - 0.82 = 0.18$.

(c) The first person on the standby list hopes that 99 or fewer people show up for the flight. Therefore, $P(x \le 99) = 0.05 + 0.10 + 0.12 + 0.14 + 0.24 = 0.65$. The third person on the standby list hopes that 97 or fewer people show up for the flight. Therefore, $P(x \le 97) = 0.05 + 0.10 + 0.12 = 0.27$.

6.105: (a) $P(x < 10) = (0.05)(10) = 0.5$; $P(x > 15) = (0.05)(20 - 15) = 0.25$

(b) $P(7 < x < 12) = (0.05)(12 - 7) = 0.25$

(c) $P(x < c) = 0.9$ means that the area under the density curve between 0 and c is 0.9. Therefore, $(0.05)c = 0.9$, or $c = 0.9 / 0.05 = 18$ minutes.

6.107: (a) Supplier 1

(b) Supplier 2

(c) I would recommend supplier 1 because the bulbs will last, on average, longer than those from supplier 2. Additionally, the bulbs from supplier 1 have less variability in their lifetimes, so there is more consistency in the bulb lifetimes when compared with supplier 2.

(d) Approximately 1000 hours.

(e) Approximately 100 hours.

6.109: Let x represent the left atrial diameter.

(a) $P(x < 24) = P\left(z < \frac{24 - 26.4}{4.2}\right) = P(z < -0.571) = 0.284$

(b) $P(x > 32) = P\left(z > \frac{32 - 26.4}{4.2}\right) = P(z > 1.333) = 1 - P(z \le 1.333) = 1 - 0.909 = 0.091$

(c) $P(25 < x < 30) = P\left(\frac{25 - 26.4}{4.2} < z < \frac{30 - 26.4}{4.2}\right) = P(-0.333 < z < 0.857) = 0.435$

(d) We need the value of z^* such that $P(z > z^*) = 0.20$, or $P(z \le z^*) = 0.80$. Using technology, $z^* = 0.8416$. Therefore,

$0.8416 = \frac{x - 26.4}{4.2} \Rightarrow x = (0.8416)(4.2) + 26.4 = 29.93$. 29.93 mm is the value for which only about 20% have a larger left atrial diameter.

6.111: (a) $P(x > 50) = 1 - P(x \le 50) = 1 - P\left(z \le \frac{50 - 45}{5}\right) = 1 - P(z \le 1) = 1 - 0.841 = 0.159$

(b) We need the value of z^* such that $P(z \le z^*) = 0.90$. From the standard normal table or using technology, $z^* = 1.28$. Therefore,

$1.28 = \frac{x - 45}{5} \Rightarrow x = (1.28)(5) + 45 = 51.4$ minutes. 51.4 minutes should be allowed for 90% of students taking the test to be able to finish in the allotted time.

(c) We need the value of z^* such that $P(z \le z^*) = 0.25$. From the standard normal table or using technology, $z^* = -0.674$. Therefore,

$-0.674 = \frac{x-45}{5} \Rightarrow x = (-0.674)(5) + 45 = 41.6$ minutes. 41.6 minutes is required for the fastest 25% of all students to complete the exam.

6.113: Yes, the data suggest that the cadmium concentration distribution is not normal because of the curvature of the points apparent in the normal probability plot.

Chapter 7 An Overview of Statistical Inference—Learning from Data

Section 7.1 Exercise Set 1

7.1: The inferences made are ones that involve estimation.

7.2: (a) American teenagers between the ages of 12 and 17.

(b) The percentage of teens that own a cell phone, the percentage of teens that use a cell phone to send and receive text messages, and the percentage of teens age 16 – 17 who have used a cell phone to text while driving.

(c) No, the actual percentage of teens owning a cell phone is not necessarily exactly 75%. The computed statistic, 75%, was based on one sample. Another sample might yield another value of the statistic, and since the statistic is used to estimate a population characteristic, might yield another value.

(d) I would expect the estimate of the percentage of teens who owns a cell phone to be more accurate. The sample contained teens aged 12-17, however only those teens aged 16 and 17 were asked about texting while driving. The number of teens aged 16 and 17 is a subset of the overall sample, and, as such, the estimate is based on a smaller sample. Smaller samples tend to yield less precise estimates.

7.3: The inference made is one that involves hypothesis testing.

7.4: (a) People driving along stretches of highway that have digital billboards.

(b) The time required to respond to road signs is greater when digital billboards are present.

(c) Answers will vary. In addition to the information provided, I would like to know if there was a control group in which the subjects were not subjected to digital billboards or a group in which subjects were not subjected to any change in the display on the digital billboard. I would also like to know if the subjects were randomly selected or not, and whether or not the subjects were randomly assigned to any experimental groups.

(d) The risk of an incorrect conclusion would have been lower if 100 people had participated.

7.5: (a) Answers will vary. One possible answer is: "What proportion of students approve of a recent decision made by the University to increase athletic fees in order to upgrade facilities?"

(b) Answers will vary. One possible answer is: "Do more than half of students approve of the recent decision made by the University to increase athletic fees in order to upgrade facilities?"

Additional Exercises for Section 7.1

7.11: The inference made is one that involves hypothesis testing.

Section 7.2 Exercise Set 1

7.13: Estimate a population mean. The data are numerical, rather than categorical, so we are probably interested in a mean.

7.14: The type of data determines what graphical, numerical, and inferential method or methods apply.

7.15: The other two questions are (1) **Q: Question Type** (estimation problem or hypothesis testing problem), and (2) **S: Study Type** (sample data or experiment data).

7.16: Q: Estimation

S: Sample data

T: Univariate, categorical data

N: One sample

7.17: Q: Hypothesis testing

S: Sample data

T: Univariate, numerical data

N: Two samples

7.18: Q: Hypothesis testing

S: Sample data

T: bivariate, categorical and numerical

N: One sample

Additional Exercises for Section 7.2

7.25: Q: Estimation

S: Sample data

T: Univariate, categorical

N: One sample

7.27: Q: Hypothesis testing

S: Experiment

T: Univariate, categorical

N: Two treatments

7.29: Q: Hypothesis testing

S: Experiment data

T: Univariate, numerical

N: Two treatments

Are You Ready to Move On? Chapter 7 Review Exercises

7.31: Estimation, because the researchers were interested in the mean rating given to the wine under each of the five musical conditions. The researchers did not indicate that they had a claim that they wanted to test.

7.33: (a) Answers will vary. One possible answer is "What proportion of people who purchased season tickets for home games of the New York Yankees purchased alcoholic beverages during the game?"

(b) Answers will vary. One possible answer is "Do fewer than 50% of people who purchased season tickets for home games of the New York Yankees drive to the game?"

7.35: (a) The proportion of people who take a garlic supplement who get a cold is lower than the proportion of those who do not take a garlic supplement who get a cold.

(b) Yes, it is possible that the conclusion that the proportion of people who take garlic who get a cold is lower than the proportion for those who do not take garlic is incorrect. It is possible with experiment data that the observed difference in treatment effects may be due to chance variability in the response variable and the random assignment to treatments, and not the direct result of the treatment.

(c) If the number of people participating in the study had been 50, the chance of an incorrect conclusion would be greater than the risk for the study described.

7.37: The type of data collected determines not only the type of numerical and graphical methods that can be applied, but also the particular inferential method or methods that can be used. Numerical data require different inferential methods than categorical data, and univariate data require different inferential methods than bivariate data.

7.39: Q: Hypothesis testing

S: Experiment data

T: Univariate, categorical

N: Two treatments

7.41: (a) Estimate, Method, Check, Calculate, Communicate results

(b) The only difference is that the "estimate" step is replaced with "hypotheses," where you determine the hypotheses you wish to test, rather than the population characteristic or treatment effect you wish to estimate.

Chapter 8 Sampling Variability and Sampling Distributions

Section 8.1 Exercise Set 1

8.1: I would not expect that $\hat{p}$ for a different random sample would also be equal to 0.428. That specific value for $\hat{p}$ was based on one particular sample. The concept of sample-to-sample variability tells us that a different sample would likely yield a different value of $\hat{p}$.

8.2: (a) The histogram on the left indicates that the value of $\hat{p}$ has more sample-to-sample variability than the histogram on the right. Although the two histograms are similar, the histogram on the left has values that tend to deviate more from the center than the histogram on the right, indicating a larger standard deviation of the values of $\hat{p}$.

(b) Samples of size $n = 75$ would yield a value of $\hat{p}$ that is more likely to be close to 0.55. The fact that the histogram on the right seems to have less sample-to-sample variability, is centered at 0.55, and has tall bars close to 0.55 supports this choice.

8.3: (a) The number that appears in boldface in this statement a population proportion.

(b) The correct notation is $p = 0.22$.

8.4: (a) The number that appears in boldface in this statement a sample proportion.

(b) The correct notation is $\hat{p} = 0.38$.

8.5: p is the proportion of the entire population, and $\hat{p}$ is the proportion of the sample.

Additional Exercises for Section 8.1

8.11: Sample statistics are computed from a sample. Since one sample from the population is likely to differ from other possible samples taken from the same population, the sample statistic computed from each sample is likely to be different.

8.13: (a) The number that appears in boldface in this statement a sample proportion.

(b) The correct notation is $\hat{p} = 0.45$.

Section 8.2 Exercise Set 1

8.15: (a) mean: $\mu_{\hat{p}} = 0.65$; standard deviation: $\sigma_{\hat{p}} = \sqrt{\dfrac{0.65(1-0.65)}{10}} = 0.151$

(b) mean: $\mu_{\hat{p}} = 0.65$; standard deviation: $\sigma_{\hat{p}} = \sqrt{\dfrac{0.65(1-0.65)}{20}} = 0.107$

(c) mean: $\mu_{\hat{p}} = 0.65$; standard deviation: $\sigma_{\hat{p}} = \sqrt{\dfrac{0.65(1-0.65)}{30}} = 0.087$

(d) mean: $\mu_{\hat{p}} = 0.65$; standard deviation: $\sigma_{\hat{p}} = \sqrt{\dfrac{0.65(1-0.65)}{50}} = 0.067$

(e) mean: $\mu_{\hat{p}} = 0.65$; standard deviation: $\sigma_{\hat{p}} = \sqrt{\dfrac{0.65(1-0.65)}{100}} = 0.048$

(f) mean: $\mu_{\hat{p}} = 0.65$; standard deviation: $\sigma_{\hat{p}} = \sqrt{\dfrac{0.65(1-0.65)}{200}} = 0.034$

8.16: The rule of thumb is that if both $np \geq 10$ and $n(1-p) \geq 10$, then the sampling distribution will be approximately normal.

For $p = 0.65$, the table below shows the calculations necessary to determine the sample sizes that would have approximately normal sampling distributions:

n	np	$n(1-p)$	Both np and $n(1-p)$ at least 10?	Sampling distribution approximately normal?
10	6.5	3.5	No	No
20	13	7	No	No
30	19.5	10.5	Yes	Yes
50	32.5	17.5	Yes	Yes
100	65	35	Yes	Yes
200	130	70	Yes	Yes

The samples of size $n = 30$, $n = 50$, $n = 100$, and $n = 200$ satisfy the rule of thumb for $p = 0.65$.

For $p = 0.2$, the table below shows the calculations necessary to determine the sample sizes that would have approximately normal sampling distributions:

n	np	$n(1-p)$	Both np and $n(1-p)$ at least 10?	Sampling distribution approximately normal?
10	2	8	No	No
20	4	16	No	No
30	6	24	No	No
50	10	40	Yes	Yes
100	20	80	Yes	Yes
200	40	160	Yes	Yes

The samples of size $n = 50$, $n = 100$, and $n = 200$ satisfy the rule of thumb for $p = 0.2$.

8.17: (a) mean: $\mu_{\hat{p}} = 0.07$; standard deviation: $\sigma_{\hat{p}} = \sqrt{\frac{0.07(1-0.07)}{100}} = 0.0255$

(b) The sampling distribution is not approximately normal because $np = (100)(0.07) = 7$ is less than the required 10.

(c) The change in sample size does not change the mean of the sampling distribution. However, the standard deviation will decrease to 0.0180. The mean does not change when the sample size is increased because the sampling distribution is always centered at the population value (in this case, $\mu_{\hat{p}} = 0.07$) regardless of the sample size. The standard deviation of the sampling distribution will decrease as the sample size increases because the sample size (n) is in the denominator of the formula for standard deviation. As sample size increases, standard deviation of the sampling distribution of $\hat{p}$ decreases.

(d) The sampling distribution of $\hat{p}$ is approximately normal because $np = (200)(0.07) = 14$ and $n(1-p) = (200)(1-0.07) = 186$. Both of these values are at least 10, so we have satisfied the rule of thumb.

8.18: (a) mean: $\mu_{\hat{p}} = 0.15$; standard deviation: $\sigma_{\hat{p}} = \sqrt{\frac{0.15(1-0.15)}{100}} = 0.0357$

(b) The sampling distribution is approximately normal because $np = (100)(0.15) = 15$ and $n(1-p) = (100)(1-0.15) = 85$ are both at least 10.

(c) The change in sample size does not change the mean of the sampling distribution. However, the standard deviation will decrease to 0.0252. The mean does not change when the sample size is increased because the sampling distribution is always centered at the population value (in this case, $\mu_{\hat{p}} = 0.15$) regardless of the sample size. The standard deviation of the sampling distribution will decrease as the sample size increases because the sample size (*n*) is in the denominator of the formula for standard deviation.

(d) Yes, the sampling distribution of $\hat{p}$ is approximately normal because $np = (200)(0.15) = 30$ and $n(1-p) = (200)(1-0.15) = 170$ are both at least 10.

Additional Exercises for Section 8.2

8.23: The smallest standard deviation occurs for $n = 100$ and $p = 0.5$.

8.25: For approximate normality of the sampling distribution of $\hat{p}$, both np and $n(1-p)$ must be at least 10. The tables below summarize the proportions and sample sizes, and indicate which scenarios satisfy the rule of thumb for approximate normality.

For $p = 0.2$:

n	np	$n(1-p)$	Both np and $n(1-p)$ at least 10?	Sampling distribution approximately normal?
10	2	8	No	No
25	5	20	No	No
50	10	40	Yes	Yes
100	20	80	Yes	Yes

The samples of size $n = 50$ and $n = 100$ satisfy the rule of thumb for $p = 0.2$.

For $p = 0.8$:

n	np	$n(1-p)$	Both np and $n(1-p)$ at least 10?	Sampling distribution approximately normal?
10	8	2	No	No
25	20	5	No	No
50	40	10	Yes	Yes
100	80	20	Yes	Yes

The samples of size $n = 50$ and $n = 100$ satisfy the rule of thumb for $p = 0.8$.

For $p = 0.6$:

n	np	$n(1-p)$	Both np and $n(1-p)$ at least 10?	Sampling distribution approximately normal?
10	6	4	No	No
25	15	10	Yes	Yes
50	30	20	Yes	Yes
100	60	40	Yes	Yes

The samples of size $n = 25$, $n = 50$, and $n = 100$ satisfy the rule of thumb for $p = 0.6$.

8.27: For approximate normality of the sampling distribution of $\hat{p}$, both np and $n(1-p)$ must be at least 10. The tables below summarize the proportions and sample sizes, and indicate which scenarios satisfy the rule of thumb for approximate normality.

n	p	np	$n(1-p)$	Both np and $n(1-p)$ at least 10?	Sampling distribution approximately normal?
40	0.20	8	32	No	No
40	0.45	18	22	Yes	Yes
40	0.70	28	12	Yes	Yes
40	0.90	36	4	No	No
75	0.20	15	60	Yes	Yes
75	0.45	33.75	41.25	Yes	Yes
75	0.70	52.5	22.5	Yes	Yes
75	0.90	67.5	7.5	No	No

For samples of size $n = 40$, the rule of thumb is satisfied for $p = 0.45$ and $p = 0.70$. For samples of size $n = 75$, the rule of thumb is satisfied for $p = 0.20$, $p = 0.45$, and $p = 0.70$.

Section 8.3 Exercise Set 1

8.29: (a) $\sigma_{\hat{p}} = \sqrt{\dfrac{0.48(1-0.48)}{500}} = 0.0223$

(b) More sample-to-sample variability in the sample proportions because $\sigma_{\hat{p}}$ is now greater than $\sigma_{\hat{p}}$ when $n = 500$.

(c) The sample size is smaller because, in order for $\sigma_{\hat{p}}$ to be greater, the denominator of $\sigma_{\hat{p}} = \sqrt{\frac{p(1-p)}{n}}$ must be smaller than when $n = 500$. The increased sample-to-sample variability is consistent with a smaller sample size.

8.30:

What You Know	How You Know It
The sampling distribution of $\hat{p}$ is centered at the actual (but unknown) value of the population proportion.	Rule 1 states that $\mu_{\hat{p}} = p$. This is true for random samples, and the description of the study says that the sample was selected at random.
An estimate of the standard deviation of $\hat{p}$, which describes how much the $\hat{p}$ values spread out around the population proportion p is 0.0153.	Rule 2 states that $\sigma_{\hat{p}} = \sqrt{\frac{p(1-p)}{n}}$. In this exercise, $n = 1000$. The value of p is not known. However, $\hat{p}$ provides an estimate of p that can be used to estimate the standard deviation of the sampling distribution. Specifically, $\hat{p} = 0.37$, so $\sigma_{\hat{p}} = \sqrt{\frac{0.37(1-0.37)}{1000}} = 0.0153$. This standard deviation provides information about how tightly the $\hat{p}$ values from different random samples will cluster around p.
The sampling distribution of $\hat{p}$ is approximately normal.	Rule 3 states that the sampling distribution of $\hat{p}$ is approximately normal if n is large and p is not too close to 0 or 1. Here the sample size is 1000. The sample includes 370 successes and 630 failures, which are both much greater than 10. So we conclude that the sampling distribution of $\hat{p}$ is approximately normal.

8.31: (a) $\hat{p} = \frac{18}{52} = 0.346$

(b) No, it is not reasonable to think that this estimate is within 0.05 of the actual value of the population proportion. From Rule 1, we know that the sampling distribution is centered at p. From Rule 2, the standard deviation of the sampling distribution of $\hat{p}$ is $\sigma_{\hat{p}} = \sqrt{\frac{0.346(1-0.346)}{52}} = 0.066$. By Rule 3, the sampling distribution of $\hat{p}$ is approximately normally distributed because the sample includes 18 successes and 34 failures, which are both greater than 10. For any variable described by a normal distribution, about 95% of the values are within two standard deviations of the center. Since the sampling distribution of $\hat{p}$ is approximately normal and is centered at the actual population proportion p, we now know that about 95% of all possible random samples of size $n = 52$ will produce a sample proportion that is within 2(0.066) = 0.132 of the actual value of the population proportion. This tells us that the sample estimate $\hat{p} = 0.346$ is likely to be within 0.132 of the actual proportion of all young adults with pierced tongues that have receding gums. This margin of error of 0.132 is over twice the value of 0.05, so it is not reasonable to think that this estimate is within 0.05 of the actual value of the population proportion.

8.32: From Rule 1, we know that the sampling distribution of $\hat{p}$ is centered at the population value $p = 0.61$. Rule 2 tells us that $\sigma_{\hat{p}} = \sqrt{\frac{0.61(1-0.61)}{1,500}} = 0.0126$. Finally, Rule 3 allows us to determine that the sampling distribution of $\hat{p}$ is approximately normal because $np = 1,500(0.61) = 915 \geq 10$ and $n(1-p) = 1,500(1-0.61) = 585 \geq 10$. We can use the results of Rules 1, 2, and 3 to find the probability of observing a sample proportion of 0.55 or smaller just by chance (due to sampling variability). This probability is $P(\hat{p} < 0.55) = P\left(z < \frac{0.55-0.61}{0.0126}\right) = P(z < -4.76) \approx 0$. Therefore, it seems likely that the proportion of California high school graduates who attend college the year after graduation is different from the national figure.

Section 8.3 Additional Exercises

8.37: No, it is not likely that this estimate is within 0.05 of the actual value of the population proportion. From Rule 1, we know that the sampling distribution is centered at p. From Rule 2, the standard deviation of the sampling distribution of $\hat{p}$ is $\sigma_{\hat{p}} = \sqrt{\frac{0.38(1-0.38)}{100}} = 0.049$. By Rule 3, the sampling distribution of $\hat{p}$ is approximately normally distributed because $np = 100(0.38) = 38 \geq 10$ and $n(1-p) = 100(1-0.38) = 62 \geq 10$. For any variable described by a normal distribution, about 95% of the values are within two standard deviations of the center. Since the sampling distribution of $\hat{p}$ is approximately normal and is centered at the actual population proportion p, we now know that about 95% of all possible random samples of size $n = 100$ will produce a sample proportion that is within 2(0.049) = 0.098 of the actual value of the population proportion. This tells us that a sample estimate is likely to be within 0.098 of the actual proportion of students at the college who would rent textbooks. This margin of error is nearly twice the value of 0.05, so it is unlikely that this estimate is within 0.05 of the population proportion.

8.39: By Rule 1, we know that the sampling distribution of $\hat{p}$ is centered at the unknown value of p, the true proportion of social network users who believe that it is not OK to "friend" your boss. Rule 2 says that the standard deviation of the sampling distribution of $\hat{p}$ is $\sigma_{\hat{p}} = \sqrt{\frac{p(1-p)}{n}}$. The value of p is not known. However, $\hat{p}$ provides an estimate of p that can be used to estimate the standard deviation of the sampling distribution. In this case, $\hat{p} = 0.56$, so the estimate of the standard deviation of the sampling distribution is $\sigma_{\hat{p}} = \sqrt{\frac{0.56(1-0.56)}{1,200}} = 0.014$. Finally, Rule 3 says that the sampling distribution of $\hat{p}$ is approximately normal if n is large and p is not too close to 0 or 1. Here the sample size is 1,200. The sample includes 672 successes (56% of 1,200) and 528 failures (44% of 1,200), which are both much greater than 10. Therefore, we can conclude that the sampling distribution of $\hat{p}$ is approximately normal. The probability of observing a sample proportion at least as large as what we actually observed ($\hat{p} = 0.56$) if the true value was 0.5 is equal to $P(\hat{p} \geq 0.56) = P\left(z \geq \frac{0.56-0.5}{0.014}\right) = P(z \geq 4.29) \approx 0.000$. It is highly unlikely that a sample proportion as large as 0.56 would be observed if the true value were 0.5 or less. Therefore, it seems plausible that the proportion of social network users who believe that it is not OK to "friend" your boss is greater than 0.5.

Are You Ready To Move On? Chapter 8 Review Exercises

8.41: It would not be surprising if $\hat{p}$ did not equal 0.64 for a second random sample of size 50. That specific value for $\hat{p}$ was based on one particular sample. The concept of sample-to-sample variability tells us that a different sample would likely yield a different value for $\hat{p}$.

8.43: Different samples will likely yield different values of $\hat{p}$, which is the concept of sampling variability (or sample-to-sample variability). However, there is only one true proportion that describes a particular population characteristic, p. Therefore, there is no variability in the population proportion.

8.45: (a) The number that appears in boldface in this statement is a population proportion.

(b) The correct notation is $p = 0.21$.

8.47: For approximate normality of the sampling distribution of $\hat{p}$, both np and $n(1-p)$ must be at least 10. The table below summarizes the proportions and sample sizes, and indicates which scenarios satisfy the rule of thumb for approximate normality.

n	p	np	$n(1-p)$	Both np and $n(1-p)$ at least 10?	Sampling distribution approximately normal?
10	0.70	7	3	No	No
20	0.70	14	6	No	No
30	0.70	21	9	No	No
50	0.70	35	15	Yes	Yes
100	0.70	70	30	Yes	Yes
200	0.70	140	60	Yes	Yes
10	0.30	3	7	No	No
20	0.30	6	14	No	No
30	0.30	9	21	No	No
50	0.30	15	35	Yes	Yes
100	0.30	30	70	Yes	Yes
200	0.30	60	140	Yes	Yes

For both $p = 0.70$ and $p = 0.30$, the rule of thumb is satisfied for $n = 50$, $n = 100$, and $n = 200$.

8.49: (a) Mean: $\mu_{\hat{p}} = p = 0.25$; Standard Deviation:

$$\sigma_{\hat{p}} = \sqrt{\frac{p(1-p)}{n}} = \sqrt{\frac{0.25(1-0.25)}{200}} = 0.031$$

(b) $np = 200(0.25) = 50 \geq 10$ and $n(1-p) = 200(1-0.25) = 150 \geq 10$; Since both np and $n(1-p)$ are both at least 10, we know that the sampling distribution of $\hat{p}$ is approximately normal.

(c) The change in sample size does not affect the mean of the sampling distribution of $\hat{p}$ because $\mu_{\hat{p}} = p$, regardless of sample size. However, the change in sample size does affect the standard deviation of the sampling distribution of $\hat{p}$ because $\sigma_{\hat{p}}$ depends on sample size. The new standard deviation is

$$\sigma_{\hat{p}} = \sqrt{\frac{0.25(1-0.25)}{50}} = 0.061.$$

(d) $np = 50(0.25) = 12.5 \geq 10$ and $n(1-p) = 50(1-0.25) = 37.5 \geq 10$; Since both np and $n(1-p)$ are both at least 10, we know that the sampling distribution of $\hat{p}$ is approximately normal.

8.51:

What You Know	How You Know It
The sampling distribution of $\hat{p}$ is centered at the actual (but unknown) value of the population proportion.	Rule 1 states that $\mu_{\hat{p}} = p$. This is true for random samples, and the description of the study says that the sample was selected at random.
An estimate of the standard deviation of $\hat{p}$, which describes how much the $\hat{p}$ values spread out around the population proportion p is 0.0133.	Rule 2 states that $\sigma_{\hat{p}} = \sqrt{\frac{p(1-p)}{n}}$. In this exercise, $n = 1000$. The value of p is not known. However, $\hat{p}$ provides an estimate of p that can be used to estimate the standard deviation of the sampling distribution. Specifically, $\hat{p} = 0.230$, so $\sigma_{\hat{p}} = \sqrt{\frac{0.230(1-0.230)}{1000}} = 0.0133$. This standard deviation provides information

	about how tightly the $\hat{p}$ values from different random samples will cluster around p.
The sampling distribution of $\hat{p}$ is approximately normal.	Rule 3 states that the sampling distribution of $\hat{p}$ is approximately normal if n is large and p is not too close to 0 or 1. Here the sample size is 1000. The sample includes 230 successes and 770 failures, which are both much greater than 10. So we conclude that the sampling distribution of $\hat{p}$ is approximately normal.

8.53: By Rule 1, we know that the sampling distribution of $\hat{p}$ is centered at the unknown value of p, the true proportion of adult Americans who believe rudeness is a worsening problem. Rule 2 says that the standard deviation of the sampling distribution of $\hat{p}$ is $\sigma_{\hat{p}} = \sqrt{\frac{p(1-p)}{n}}$. The value of p is not known. However, $\hat{p}$ provides an estimate of p that can be used to estimate the standard deviation of the sampling distribution. In this case, $\hat{p} = \frac{1,283}{2,013} = 0.637$, so the estimate of the standard deviation of the sampling distribution is $\sigma_{\hat{p}} = \sqrt{\frac{0.637(1-0.637)}{2,013}} = 0.011$. Finally, Rule 3 says that the sampling distribution of $\hat{p}$ is approximately normal if n is large and p is not too close to 0 or 1. Here the sample size is 2,013. The sample includes 1,283 successes and 730 failures, which are both much greater than 10. Therefore, we can conclude that the sampling distribution of $\hat{p}$ is approximately normal. The probability of observing a sample proportion at least as large as what we observed ($\hat{p} = 0.637$) if the true value was 0.5 is equal to $P(\hat{p} \geq 0.637) = P\left(z \geq \frac{0.637-0.5}{0.011}\right) = P(z \geq 12.45) \approx 0.000$. It is highly unlikely that a sample proportion as large as 0.637 would be observed if the true value were 0.5 or less. Therefore, it seems plausible that the proportion of adult Americans who believe rudeness is a worsening problem is greater than 0.5.

Chapter 9 Estimating a Population Proportion

Section 9.1 Exercise Set 1

9.1: An unbiased statistic with a smaller standard error is preferred because it is likely to result in an estimate that is closer to the actual value of the population characteristic than an unbiased statistic that has a larger standard error.

9.2: Statistics II and III are unbiased estimators of the population characteristic because they are centered at the true value of the population characteristic.

9.3: Statistic I is recommended because it has a smaller bias than Statistics II and III.

9.4: The standard error of the sample proportion $\hat{p}$ would be smaller for random samples of size $n = 200$.

9.5: (a) The formula for the standard error of $\hat{p}$ is $\sigma_{\hat{p}} = \sqrt{\frac{p(1-p)}{n}}$. The quantity $p(1-p)$ reaches a maximum value when $p = 0.5$ (because p ranges between 0 and 1, all values of $p(1-p)$ when $p \neq 0.5$ will be smaller than $p(1-p)$ when $p = 0.5$), which is why the standard error of $\hat{p}$ is greater when p is near 0.5 than when p is near 1.

(b) The standard error of $\hat{p}$ is the same when $p = 0.2$ as when $p = 0.8$ because when $p = 0.2$, $(1-p) = (1-0.2) = 0.8$. Similarly, when $p = 0.8$, $(1-p) = (1-0.8) = 0.2$. In either case, you are multiplying 0.2 and 0.8.

9.6: The estimate will tend to be closest to the actual value of p when the statistic is unbiased and has the smallest standard error of $\hat{p}$. In this case, the statistic $\hat{p}$ is an unbiased estimator of p, and the standard error is smallest when $n = 400$ and $p = 0.8$ (situation iii).

Section 9.1 Additional Exercises

9.13: A biased statistic might be chosen over an unbiased statistic if the bias is not too large, and the standard error of the biased statistic is much smaller than the standard error of the unbiased statistic. In this case, the resulting biased statistic might tend to be closer to the true value than an unbiased statistic.

9.15: The standard error of the sample proportion $\hat{p}$ would be smaller for random samples of size $n = 200$.

Section 9.2 Exercise Set 1

9.17: Statement 1: Incorrect because the value 0.0157 is the standard error of $\hat{p}$, and therefore approximately 32% of all possible values of $\hat{p}$ could differ from the value of the actual population proportion by more than 0.0157 (using properties of the normal distribution).

Statement 2: Correct

Statement 3: Incorrect because the phrase "will never differ from the value of the actual population proportion" is wrong. The value 0.0307 is the margin of error and indicates that in 95% of all possible random samples, the estimation error will be less than the margin of error. In 5% of the random samples, the estimation error will be greater than the margin of error.

9.18: (a) $\sigma_{\hat{p}} = \sqrt{\dfrac{p(1-p)}{n}} = \sqrt{\dfrac{0.4(1-0.4)}{100}} = 0.049$

(b) The standard error of $\hat{p}$ would be larger for a sample of size 100 because, with the sample size n in the denominator of the formula for standard error, a smaller value in the denominator would produce a larger value for the fraction.

(c) If the sample size doubles from 100 to 200, the standard error decreases by a factor of $1/\sqrt{2} \approx 0.707$.

9.19: (a) $\hat{p} = \dfrac{137}{526} = 0.260$; the sample proportion ($\hat{p}$) is the statistic that was used.

(b) $\sigma_{\hat{p}} = \sqrt{\dfrac{0.260(1-0.260)}{526}} = 0.019$

(c) $\text{margin of error} = 1.96\sqrt{\dfrac{0.260(1-0.260)}{526}} = 1.96(0.019) = 0.037$; This estimate of the proportion of all businesses that have fired workers for misuse of the Internet is unlikely to differ from the actual population proportion by more than 0.037.

9.20: (a) $n\hat{p} = 50(0.3) = 15$ and $n(1-\hat{p}) = 50(1-0.3) = 35$; both $n\hat{p}$ and $n(1-\hat{p})$ are at least 10, so the sample size is large enough for the use of the formula.

(b) $n\hat{p} = 50(0.05) = 2.5$ and $n(1-\hat{p}) = 50(1-0.05) = 47.5$; $n\hat{p}$ is not at least 10, so the use of the formula is not appropriate.

(c) $n\hat{p} = 15(0.45) = 6.75$ and $n(1-\hat{p}) = 15(1-0.45) = 8.25$; both $n\hat{p}$ and $n(1-\hat{p})$ are less than 10, so the sample size is not large enough for the use of the formula.

(d) $n\hat{p} = 100(0.01) = 1$ and $n(1-\hat{p}) = 100(1-0.01) = 99$; $n\hat{p}$ is not at least 10, so the use of the formula is not appropriate.

9.21: (a) $\hat{p} = \dfrac{2{,}998}{7{,}421} = 0.404$

(b) The sample was selected in such a way that makes it representative of the population U.S. college students. Additionally, there are 2,998 successes and 4,423 failures in the sample, which are both at least 10. These conditions together verify that the margin of error formula is appropriate.

(c) margin of error $= 1.96\sqrt{\dfrac{0.404(1-0.404)}{7{,}421}} = 0.011$

(d) It is unlikely that the estimated proportion of U.S. college students who use the Internet more than 3 hours per day ($\hat{p} = 0.404$) will differ from the true proportion by more than 0.011.

9.22: (a) $\hat{p} = \dfrac{1{,}720}{6{,}212} = 0.277$

(b) The sample is a random sample from the population of American children. Additionally, there are 1,720 successes and 4492 failures in the sample, which are both greater than 10. These conditions together verify that the margin of error formula is appropriate.

(c) margin of error $= 1.96\sqrt{\dfrac{0.277(1-0.277)}{6{,}212}} = 0.011$

(d) It is unlikely that the estimated proportion of American children who indicated that they eat fast food on a typical day ($\hat{p} = 0.277$) will differ from the true proportion by more than 0.011.

Section 9.2 Additional Exercises

9.29: margin of error $= 1.96\sqrt{\dfrac{0.82(1-0.82)}{1,002}} = 0.024$; It is unlikely that the estimated proportion of adults who believe that the shows are mostly or totally made up ($\hat{p} = 0.82$) will differ from the true proportion by more than 0.024.

9.31: In this case, margin of error $= 1.96\sqrt{\dfrac{0.25(1-0.25)}{1,002}} = 0.027$, which rounds to 3%.

9.33: In this case, margin of error $= 1.96\sqrt{\dfrac{0.43(1-0.43)}{930}} = 0.032$, which rounds to 3%.

The margin of error tells us that it is unlikely that the estimate sample proportion of all residents who feel that their financial situation has improved ($\hat{p} = 0.43$) will differ from the true proportion by more than 0.03.

Section 9.3 Exercise Set 1

9.34: (a) Recall that confidence intervals are centered at $\hat{p}$. In this case, the intervals are not centered in the same place because two different samples were taken, each yielding a different value of $\hat{p}$.

(b) Interval 2 conveys more precise information about the value of the population proportion because Interval 2 is narrower than Interval 1.

(c) A smaller sample size produces a larger margin of error. In this case, Interval 1 (being wider than Interval 2) was based on the smaller sample size.

(d) Interval 1 would have the higher confidence level, because the *z* critical value for higher confidence is larger, resulting in a wider confidence interval.

9.35: (a) The 95% confidence level would result in a wider large-sample confidence interval because the *z* critical value is larger for a higher confidence level.

(b) The sample of size 100 would result in a wider large-sample confidence interval because the standard error of $\hat{p}$ would be larger than it would for a sample of size 400.

9.36: The method used to construct this interval estimate is successful in capturing the actual value of the population proportion 95% of the time. Alternatively, 95% of all possible interval estimates for this sample size taken from the same population would capture the actual value of the population proportion.

9.37: (a) $n\hat{p} = 50(0.3) = 15$ and $n(1-\hat{p}) = 50(1-0.3) = 35$; both $n\hat{p}$ and $n(1-\hat{p})$ are at least 10, so the sample size is large enough for the use of the formula.

(b) $n\hat{p} = 50(0.05) = 2.5$ and $n(1-\hat{p}) = 50(1-0.05) = 47.5$; $n\hat{p}$ is not at least 10, so the use of the formula is not appropriate.

(c) $n\hat{p} = 15(0.45) = 6.75$ and $n(1-\hat{p}) = 15(1-0.45) = 8.25$; both $n\hat{p}$ and $n(1-\hat{p})$ are less than 10, so the sample size is not large enough for the appropriate use of the formula.

(d) $n\hat{p} = 100(0.01) = 1$ and $n(1-\hat{p}) = 100(1-0.01) = 99$; $n\hat{p}$ is not at least 10, so the use of the formula is not appropriate.

9.38: (a) z critical value for 90% confidence level is 1.645

(b) z critical value for 99% confidence level is 2.58

(c) z critical value for 80% confidence level is 1.28

9.39: Question type (Q): Estimation

Study type (S): Sample data

Type of data (T): One categorical variable

Number of samples or treatments (N): One sample

This particular combination of answers to QSTN confirms that a confidence interval for a population proportion is appropriate.

9.40: Estimate (E): The proportion of hiring managers and human resources professionals who use social networking sites to research job applicants, p, will be estimated.

Method (M): Because the answers to the four key questions are (Q) estimation, (S) sample data, (T) one categorical variable, and (N) one sample (see Exercise 9.39), we can construct and interpret a 95% confidence interval for the proportion of hiring managers and human resource professionals who use social networking sites to research job applicants.

Check (C): We are told that the sample is representative of hiring managers and human resources professionals. In addition, the sample includes 1200 successes and 1467 failures, which are both greater than 10. The two required conditions are satisfied.

Calculations (C): $\hat{p} = \frac{1200}{2667} = 0.45$

$$\hat{p} \pm (z \text{ critical value})\sqrt{\frac{\hat{p}(1-\hat{p})}{n}}$$
$$0.45 \pm 1.96\sqrt{\frac{0.45(1-0.45)}{2667}}$$
$$0.45 \pm 1.96(0.0096)$$
$$0.45 \pm 0.0189$$
$$(0.4311, 0.4689)$$

Communicate Results (C):

Interpret confidence interval: We are 95% confident that the actual proportion of hiring managers and human resources professionals is somewhere between 0.4311 and 0.4689.

Interpret confidence level: The method used to construct this interval estimate is successful in capturing the actual value of the population proportion 95% of the time.

9.41: (a) Using the 5-step process (EMC^3):

Estimate (E): The proportion of college freshmen who carry a credit card balance, p, will be estimated.

Method (M): Because the answers to the four key questions are (Q) estimation, (S) sample data, (T) one categorical variable, and (N) one sample, we can construct and interpret a 90% confidence interval for the proportion of college freshmen who carry a credit card balance.

Check (C): We are told that a random sample of 1,000 college freshmen was taken. In addition, $n\hat{p} = 1000(0.37) = 370 \geq 10$ and $n(1-\hat{p}) = 1000(1-0.37) = 630 \geq 10$, which are both at least 10. The two required conditions are satisfied.

Calculations (C):

$$\hat{p} \pm (z \text{ critical value})\sqrt{\frac{\hat{p}(1-\hat{p})}{n}}$$

$$0.37 \pm 1.645\sqrt{\frac{0.37(1-0.37)}{1000}}$$

$$0.37 \pm 1.645(0.0153)$$

$$0.37 \pm 0.0251$$

$$(0.3449, 0.3951)$$

Communicate Results (C):

Interpret confidence interval: We are 90% confident that the true proportion of college freshmen who carry a credit card balance is somewhere between 0.3449 and 0.3951.

Interpret confidence level: The method used to construct this interval estimate is successful in capturing the actual value of the population proportion 90% of the time.

(b) Using the 5-step process (EMC^3):

Estimate (E): The proportion of college seniors who carry a credit card balance, p, will be estimated.

Method (M): Because the answers to the four key questions are (Q) estimation, (S) sample data, (T) one categorical variable, and (N) one sample, we can construct and interpret a 90% confidence interval for the proportion of college seniors who carry a credit card balance.

Check (C): We are told that a random sample of 1,000 college seniors was taken. In addition, $n\hat{p} = 1000(0.48) = 480 \geq 10$ and $n(1-\hat{p}) = 1000(1-0.48) = 520 \geq 10$, which are both at least 10. The two required conditions are satisfied.

Calculations (C):

$$\hat{p} \pm (z \text{ critical value})\sqrt{\frac{\hat{p}(1-\hat{p})}{n}}$$

$$0.48 \pm 1.645\sqrt{\frac{0.48(1-0.48)}{1000}}$$

$$0.48 \pm 1.645(0.0158)$$

$$0.48 \pm 0.0260$$

$$(0.454, 0.506)$$

Communicate Results (C):

Interpret confidence interval: We are 90% confident that the true proportion of college seniors who carry a credit card balance is somewhere between 0.454 and 0.506.

Interpret confidence level: The method used to construct this interval estimate is successful in capturing the actual value of the population proportion 90% of the time.

(c) The two confidence intervals from (a) and (b) do not have the same width because the standard errors, and hence the margin of error, are based on two different values for $\hat{p}$.

9.42: (a) Using the 5-step process (EMC^3):

Estimate (E): The proportion of U.S. adults for whom math was their favorite subject, p, will be estimated.

Method (M): Because the answers to the four key questions are (Q) estimation, (S) sample data, (T) one categorical variable, and (N) one sample, we can construct and interpret a 95% confidence interval for the proportion of U.S. adults for whom math was their favorite subject.

Check (C): We are told that a random sample of 1,000 U.S. adults was taken. In addition, the sample contains 230 successes and 770 failures, which are both at least 10. The two required conditions are satisfied.

Calculations (C): $\hat{p} = \dfrac{230}{1000} = 0.23$

$$\hat{p} \pm (z \text{ critical value})\sqrt{\frac{\hat{p}(1-\hat{p})}{n}}$$

$$0.23 \pm 1.96\sqrt{\frac{0.23(1-0.23)}{1000}}$$

$$0.23 \pm 1.96(0.0133)$$

$$0.23 \pm 0.0261$$

$$(0.2039, 0.2561)$$

Communicate Results (C):

Interpret confidence interval: We are 95% confident that the true proportion of U.S. adults for whom math was their most favorite subject is somewhere between 0.2039 and 0.2561.

Interpret confidence level: The method used to construct this interval estimate is successful in capturing the actual value of the population proportion 95% of the time.

(b) Using the 5-step process (EMC^3):

Estimate (E): The proportion of U.S. adults for whom math was their least favorite subject, p, will be estimated.

Method (M): Because the answers to the four key questions are (Q) estimation, (S) sample data, (T) one categorical variable, and (N) one sample, we can construct and interpret a 95% confidence interval for the proportion of U.S. adults for whom math was their least favorite subject.

Check (C): We are told that a random sample of 1,000 U.S. adults was taken. In addition, the sample contains 370 successes and 630 failures, which are both at least 10. The two required conditions are satisfied.

Calculations (C): $\hat{p} = \dfrac{370}{1000} = 0.37$

$$\hat{p} \pm (z \text{ critical value})\sqrt{\frac{\hat{p}(1-\hat{p})}{n}}$$

$$0.37 \pm 1.96\sqrt{\frac{0.37(1-0.37)}{1000}}$$

$$0.37 \pm 1.96(0.0153)$$

$$0.37 \pm 0.0299$$

$$(0.3401, 0.3999)$$

Communicate Results (C):

Interpret confidence interval: We are 95% confident that the true proportion of U.S. adults for whom math was their least favorite subject is somewhere between 0.3401 and 0.3999.

Interpret confidence level: The method used to construct this interval estimate is successful in capturing the actual value of the population proportion 95% of the time.

9.43: (a) Using the 5-step process (EMC^3):

Estimate (E): The proportion of all adult Americans who planned to purchase a Valentine's Day gift for their pet, p, will be estimated.

Method (M): Because the answers to the four key questions are (Q) estimation, (S) sample data, (T) one categorical variable, and (N) one sample, we can construct and interpret a 95% confidence interval for the proportion of all adult Americans who planned to purchase a Valentine's Day gift for their pet.

Check (C): We are told that a random sample of 200 U.S. adults was taken. In addition, $n\hat{p} = 200(0.173) = 34.6 \geq 10$ and $n(1-\hat{p}) = 200(1-0.173) = 165.4 \geq 10$. The two required conditions are satisfied.

Calculations (C):

$$\hat{p} \pm (z \text{ critical value})\sqrt{\frac{\hat{p}(1-\hat{p})}{n}}$$

$$0.173 \pm 1.96\sqrt{\frac{0.173(1-0.173)}{200}}$$

$$0.173 \pm 1.96(0.0267)$$

$$0.173 \pm 0.0524$$

$$(0.1206, 0.2254)$$

Communicate Results (C):

Interpret confidence interval: We are 95% confident that the true proportion of all adult Americans who planned to purchase a Valentine's Day gift for their pet is somewhere between 0.1206 and 0.2254.

Interpret confidence level: The method used to construct this interval estimate is successful in capturing the actual value of the population proportion 95% of the time.

(b) The 95% confidence interval computed for the actual sample size would have been narrower than the confidence interval computed in part (a) because the standard error, and hence the margin of error, would have been smaller.

Section 9.3 Additional Exercises

9.55: (a) Using the 5-step process (EMC^3):

Estimate (E): The proportion of all adult Americans who would say that lying is never justified, p, will be estimated.

Method (M): Because the answers to the four key questions are (Q) estimation, (S) sample data, (T) one categorical variable, and (N) one sample, we can construct and interpret a 90% confidence interval for the proportion of all adult Americans who would say that lying is never justified.

Check (C): We are told that a random sample of 1,000 U.S. adults was taken. In addition, $n\hat{p} = 1000(0.52) = 520 \geq 10$ and $n(1-\hat{p}) = 1000(1-0.52) = 480 \geq 10$. The two required conditions are satisfied.

Calculations (C):

$$\hat{p} \pm (z \text{ critical value})\sqrt{\frac{\hat{p}(1-\hat{p})}{n}}$$

$$0.52 \pm 1.645\sqrt{\frac{0.52(1-0.52)}{1000}}$$

$$0.52 \pm 1.645(0.0158)$$

$$0.52 \pm 0.0260$$

$$(0.494, 0.546)$$

Communicate Results (C):

Interpret confidence interval: We are 90% confident that the true proportion of adult Americans who would say that lying is never justified is somewhere between 0.494 and 0.546.

Interpret confidence level: The method used to construct this interval estimate is successful in capturing the actual value of the population proportion 90% of the time.

(b) Using the 5-step process (EMC^3):

Estimate (E): The proportion of all adult Americans who would say that it is often or sometimes OK to lie to avoid hurting someone's feelings, p, will be estimated.

Method (M): Because the answers to the four key questions are (Q) estimation, (S) sample data, (T) one categorical variable, and (N) one sample, we can construct and interpret a 90% confidence interval for the proportion of all adult Americans who would say that it is often or sometimes OK to lie to avoid hurting someone's feelings.

Check (C): We are told that a random sample of 1,000 U.S. adults was taken. In addition, the sample contains 650 successes and 350 failures, which are both at least 10. The two required conditions are satisfied.

Calculations (C): $\hat{p} = \frac{650}{1000} = 0.65$

$$\hat{p} \pm (z \text{ critical value})\sqrt{\frac{\hat{p}(1-\hat{p})}{n}}$$

$$0.65 \pm 1.645\sqrt{\frac{0.65(1-0.65)}{1000}}$$

$$0.65 \pm 1.645(0.0151)$$

$$0.65 \pm 0.0248$$

$$(0.6252, 0.6748)$$

Communicate Results (C):

Interpret confidence interval: We are 90% confident that the true proportion of adult Americans who would say that it is often or sometimes OK to lie to avoid hurting someone's feelings is somewhere between 0.6252 and 0.6748.

Interpret confidence level: The method used to construct this interval estimate is successful in capturing the actual value of the population proportion 90% of the time.

(c) The confidence interval in part (a) indicates that it is plausible that at least 50% of adult Americans would say that lying is never justified, and the confidence interval in part (b) indicates that it is also plausible that well over 50% of adult Americans would say that it is often or sometimes OK to lie to avoid hurting someone's feelings. These are contradictory responses. In addition, there are 520 respondents who said that lying is never justified, and 650 who said that it is often or sometimes OK to lie to avoid hurting someone's feelings. These two groups have a combined number of responses that is greater than 1,000 (520 + 650 = 1,170). Therefore, some people said that lying is never justified and that it is often or sometimes OK to lie to avoid hurting someone's feelings. These are contradictory responses.

9.57: Using the 5-step process (EMC^3):

Estimate (E): The proportion of all Australian children who would say that they watch TV before school, p, will be estimated.

Method (M): Because the answers to the four key questions are (Q) estimation, (S) sample data, (T) one categorical variable, and (N) one sample, we can construct and interpret a 95% confidence interval for the proportion of all Australian children who would say that they watch TV before school.

Check (C): We are told that the sample contains 1,710 schoolchildren in Australia. There is no indication whether or not the sample was randomly selected, or is representative of the population of Australian schoolchildren. In addition, the sample contains 1,060 successes and 650 failures, which are both at least 10. It is unclear whether or not the two required conditions are satisfied, so proceed with caution (see comment below).

Calculations (C): $\hat{p} = \frac{1,060}{1,710} = 0.620$

$$\hat{p} \pm (z \text{ critical value})\sqrt{\frac{\hat{p}(1-\hat{p})}{n}}$$

$$0.620 \pm 1.96\sqrt{\frac{0.620(1-0.620)}{1710}}$$

$$0.620 \pm 1.96(0.0117)$$

$$0.620 \pm 0.0230$$

$$(0.597, 0.643)$$

Communicate Results (C):

Interpret confidence interval: We are 95% confident that the true proportion of all Australian children who would say that they watch TV before school is between 0.597 and 0.643.

Interpret confidence level: The method used to construct this interval estimate is successful in capturing the actual value of the population proportion 95% of the time.

In order for the method used to construct the interval to be valid, the sample must have either been randomly selected from the population of interest, or the sample must have been selected in such a way that it should result in a representative sample from the population.

9.59: Using the 5-step process (EMC^3):

Estimate (E): The proportion of all full-time workers so angered in the last year that they wanted to hit a coworker, p, will be estimated.

Method (M): Because the answers to the four key questions are (Q) estimation, (S) sample data, (T) one categorical variable, and (N) one sample, we can construct and interpret a 90% confidence interval for the proportion of all full-time workers so angered in the last year that they wanted to hit a coworker.

Check (C): We are told that it is reasonable to regard the sample is representative of the population of full-time workers. In addition, the sample contains 125 successes and 625 failures, which are both at least 10. The two required conditions are satisfied.

Calculations (C): $\hat{p} = \frac{125}{750} = 0.167$

$$\hat{p} \pm (z \text{ critical value})\sqrt{\frac{\hat{p}(1-\hat{p})}{n}}$$

$$0.167 \pm 1.645\sqrt{\frac{0.167(1-0.167)}{750}}$$

$$0.167 \pm 1.645(0.0136)$$

$$0.167 \pm 0.0224$$

$$(0.1446, 0.1894)$$

Communicate Results (C):

Interpret confidence interval: We are 90% confident that the true proportion of all full-time workers so angered in the past year that they wanted to hit a coworker is between 0.1446 and 0.1894.

Interpret confidence level: The method used to construct this interval estimate is successful in capturing the actual value of the population proportion 90% of the time.

Section 9.4 Exercise Set 1

9.61: Assuming a 95% confidence level, and using a conservative estimate of $p = 0.5$,

$n = p(1-p)\left(\frac{1.96}{M}\right)^2 = 0.5(1-0.5)\left(\frac{1.96}{0.02}\right)^2 = 2401$. The sample should include 2,401 students.

9.62: Assuming a 95% confidence level, and using the preliminary estimate of $p = 0.27$,

$$n = p(1-p)\left(\frac{1.96}{M}\right)^2 = 0.27(1-0.27)\left(\frac{1.96}{0.05}\right)^2 = 302.872$$. The sample should contain 303 individuals. Using the conservative estimate of $p = 0.5$,

$$n = p(1-p)\left(\frac{1.96}{M}\right)^2 = 0.5(1-0.5)\left(\frac{1.96}{0.05}\right)^2 = 384.16$$. The sample should contain 385 individuals. As expected, the sample size computed using the conservative estimate is larger than the sample size computed using the preliminary estimate. The sample size 385 should be used for this study because it will guarantee a margin of error of no greater than 0.05 because the margin of error is largest with $p = 0.5$. The smaller sample size (303) will only guarantee a margin of error no greater than 0.05 if $p < 0.27$ or if $p > 1 - 0.27 = 0.73$. If $\hat{p}$ lies between 0.27 and 0.73, which could happen with a different sample, then the margin of error would be greater than 0.05.

9.63: Assuming a 95% confidence level, and using a conservative estimate of $p = 0.5$,

$$n = p(1-p)\left(\frac{1.96}{M}\right)^2 = 0.5(1-0.5)\left(\frac{1.96}{0.1}\right)^2 = 96.04$$. The sample should include 97 handles.

Section 9.4 Additional Exercises

9.67: Assuming a 95% confidence level, and using the preliminary estimate of $p = 0.32$,

$$n = p(1-p)\left(\frac{1.96}{M}\right)^2 = 0.32(1-0.32)\left(\frac{1.96}{0.05}\right)^2 = 334.373$$. The sample should contain 335 individuals. Using the conservative estimate of $p = 0.5$,

$$n = p(1-p)\left(\frac{1.96}{M}\right)^2 = 0.5(0.5)\left(\frac{1.96}{0.05}\right)^2 = 384.16$$. The sample should contain 385 individuals. As expected, the sample size computed using the conservative estimate is larger than the sample size computed using the preliminary estimate. The sample size 385 should be used for this study because it will guarantee a margin of error of no greater than 0.05 because the margin of error is largest with $p = 0.5$. The smaller sample size will only guarantee a margin of error no greater than 0.05 if $p < 0.32$ or if $p > 1 - 0.32 = 0.68$. If $\hat{p}$ lies between 0.32 and 0.68, which could happen with a different sample, then the margin of error would be greater than 0.05.

9.69: (a) Using the 5-step process (EMC^3):

Estimate (E): The proportion of all such patients under 50 years old who experience a failure within the first 2 years, *p*, will be estimated.

Method (M): Because the answers to the four key questions are (Q) estimation, (S) sample data, (T) one categorical variable, and (N) one sample, we can construct and interpret a 95% confidence interval for the proportion of all such patients under 50 years old who experience a failure within the first 2 years.

Check (C): We are told that it is reasonable to regard the sample as representative of patients who receive this type of defibrillator. In addition, the sample contains 18 successes and 71 failures, which are both at least 10. The two required conditions are satisfied.

Calculations (C): $\hat{p} = \frac{18}{89} = 0.202$

$$\hat{p} \pm (z \text{ critical value}) \sqrt{\frac{\hat{p}(1-\hat{p})}{n}}$$

$$0.202 \pm 1.96 \sqrt{\frac{0.202(1-0.202)}{89}}$$

$$0.202 \pm 1.96(0.0426)$$

$$0.202 \pm 0.0834$$

$$(0.1186, 0.2854)$$

Communicate Results (C):

Interpret confidence interval: We are 95% confident that the true proportion of all such patients under 50 years old who experience a failure within the first 2 years is between 0.1186 and 0.2854.

Interpret confidence level: The method used to construct this interval estimate is successful in capturing the actual value of the population proportion 95% of the time.

(b) Using the 5-step process (EMC^3):

Estimate (E): The proportion of all such patients age 50 and older who experience a failure within the first 2 years, p, will be estimated.

Method (M): Because the answers to the four key questions are (Q) estimation, (S) sample data, (T) one categorical variable, and (N) one sample, we can construct and interpret a 99% confidence interval for the proportion of all such patients age 50 and older who experience a failure within the first 2 years.

Check (C): We are told that it is reasonable to regard the sample as representative of patients who receive this type of defibrillator. In addition, the sample contains 13 successes and 349 failures, which are both at least 10. The two required conditions are satisfied.

Calculations (C): $\hat{p} = \frac{13}{362} = 0.0359$

$$\hat{p} \pm (z \text{ critical value})\sqrt{\frac{\hat{p}(1-\hat{p})}{n}}$$

$$0.0359 \pm 2.58\sqrt{\frac{0.0359(1-0.0359)}{362}}$$

$$0.0359 \pm 2.58(0.00978)$$

$$0.0359 \pm 0.02523$$

$$(0.0107, 0.0611)$$

Communicate Results (C):

Interpret confidence interval: We are 99% confident that the true proportion of all such patients age 50 or older who experience a failure within the first 2 years is between 0.0107 and 0.0611.

Interpret confidence level: The method used to construct this interval estimate is successful in capturing the actual value of the population proportion 99% of the time.

(c) For a 95% confidence level, and using a preliminary estimate of $p = 0.202$,

$n = p(1-p)\left(\frac{1.96}{M}\right)^2 = 0.202(1-0.202)\left(\frac{1.96}{0.03}\right)^2 = 688.056$. The sample should include 689 individuals.

Are You Ready To Move On? Chapter 9 Review Exercises

9.71: (a) Statistic II is unbiased because the mean of the sampling distribution is equal to the actual value of the population characteristic.

(b) Statistic I is recommended as the better estimator of the population characteristic because it has a smaller standard deviation and is therefore likely to produce an estimate that is closer to 54 than Statistic II would.

9.73: The standard error of the sampling distribution when $p = 0.4$ is

$\sigma_{\hat{p}} = \sqrt{\frac{p(1-p)}{n}} = \sqrt{\frac{0.4(1-0.4)}{400}} = 0.024$, and the standard error of the sampling distribution when $p = 0.7$ is $\sigma_{\hat{p}} = \sqrt{\frac{p(1-p)}{n}} = \sqrt{\frac{0.7(1-0.7)}{400}} = 0.023$. Therefore, $\hat{p}$ from a random sample of size 400 tends to be closer to the actual value when $p = 0.7$. Recall as well that when $p = 0.5$, the standard error of the sampling distribution is at a maximum, and decreases as p moves away from $p = 0.5$, toward 0 or 1. Since $p = 0.4$ is closer to $p = 0.5$ than $p = 0.7$ is, the smaller standard error will occur when $p = 0.7$.

9.75: Statement 1: Correct

Statement 2: Incorrect. 95% of all possible sample proportions computed from samples of size 500 collected in an identical manner from the population will be within 0.0429 of the actual population proportion (this is the margin of error for a 95% confidence level). 5% of the sample proportions would differ from the actual population proportion by more than 0.0429.

Statement 3: Incorrect. 68% of all possible sample proportions computed from samples of size 500 collected in an identical manner from the population will be within 0.0219 of the actual population proportion (this is the standard deviation of the sampling distribution, which is equivalent to the margin of error for 68% confidence level). 32% of the sample proportions would differ from the actual population proportion by more than 0.0219.

9.77: (a) $\hat{p} = \frac{722}{4013} = 0.18$

(b) The two conditions are

(1) large sample size: The number of successes and failures in the sample are at least 10. In this case, there are 722 successes, and 3,291 failures, which are both much greater than 10.

(2) random selection of the sample, or the sample is representative of the population: We are told that the sample is representative of the population.

(c) margin of error $= 1.96\sqrt{\frac{0.18(1-0.18)}{4013}} = 0.0119$

(d) It is unlikely that the estimated proportion of all adult Americans who have seen a ghost, $\hat{p} = 0.18$, differs from the value of the actual population proportion of all adult Americans who have seen a ghost by more than 0.0119.

(e)

$$\hat{p} \pm (z \text{ critical value})\sqrt{\frac{\hat{p}(1-\hat{p})}{n}}$$

$$0.18 \pm 1.645\sqrt{\frac{0.18(1-0.18)}{4013}}$$

$$0.18 \pm 1.645(0.006065)$$

$$0.18 \pm 0.00998$$

$$(0.17, 0.19)$$

We are 90% confident that the actual proportion of all adult Americans who have seen a ghost is somewhere between 0.17 and 0.19.

(f) A 99% confidence interval would be wider because the z critical value for 99% confidence ($z = 2.58$) is larger than the z critical value for 90% confidence ($z = 1.645$).

9.79: (a) As the confidence level increases, the width of the confidence interval also increases.

(b) As the sample size increases, the width of the confidence interval decreases.

(c) When $\hat{p} = 0.5$, the margin of error (for a fixed sample size) is maximum, and decreases symmetrically as $\hat{p}$ decreases toward 0 or increases toward 1. A larger margin of error results in a wider confidence interval.

9.81: Question type (Q): Estimation

Study type (S): Sample data

Type of data (T): One categorical variable

Number of samples or treatments (N): One sample

This particular combination of answers to QSTN confirms that a confidence interval for a population proportion is appropriate.

9.83: (a) Using the 5-step process (EMC^3):

Estimate (E): The proportion of all Americans ages 8 to 18 who own a cell phone, p, will be estimated.

Method (M): Because the answers to the four key questions are (Q) estimation, (S) sample data, (T) one categorical variable, and (N) one sample, we can construct and interpret a 90% confidence interval for the proportion of all Americans ages 8 to 18 who own a cell phone.

Check (C): We are told that it is reasonable to regard the sample as representative of Americans in this age group. In addition, the sample contains 1,321 successes and 681 failures, which are both at least 10. The two required conditions are satisfied.

Calculations (C): $\hat{p} = \dfrac{1321}{2002} = 0.66$

$$\hat{p} \pm (z \text{ critical value})\sqrt{\frac{\hat{p}(1-\hat{p})}{n}}$$

$$0.66 \pm 1.645\sqrt{\frac{0.66(1-0.66)}{2002}}$$

$$0.66 \pm 1.645(0.0106)$$

$$0.66 \pm 0.0174$$

$$(0.6426, 0.6774)$$

Communicate Results (C):

Interpret confidence interval: We are 90% confident that the true proportion of all Americans ages 8 to 18 who own a cell phone is somewhere between 0.6426 and 0.6774.

Interpret confidence level: The method used to construct this interval estimate is successful in capturing the actual value of the population proportion 90% of the time.

(b) Using the 5-step process (EMC^3):

Estimate (E): The proportion of all Americans ages 8 to 18 who own an MP3 music player, p, will be estimated.

Method (M): Because the answers to the four key questions are (Q) estimation, (S) sample data, (T) one categorical variable, and (N) one sample, we can construct and interpret a 90% confidence interval for the proportion of all Americans ages 8 to 18 who own MP3 music player.

Check (C): We are told that it is reasonable to regard the sample as representative of Americans in this age group. In addition, the sample contains 1,522 successes and 480 failures, which are both at least 10. The two required conditions are satisfied.

Calculations (C): $\hat{p} = \frac{1522}{2002} = 0.76$

$$\hat{p} \pm (z \text{ critical value})\sqrt{\frac{\hat{p}(1-\hat{p})}{n}}$$

$$0.76 \pm 1.645\sqrt{\frac{0.76(1-0.76)}{2002}}$$

$$0.76 \pm 1.645(0.0095)$$

$$0.76 \pm 0.0157$$

$$(0.7443, 0.7757)$$

Communicate Results (C):

Interpret confidence interval: We are 90% confident that the true proportion of all Americans ages 8 to 18 who own an MP3 music player is somewhere between 0.7443 and 0.7757.

Interpret confidence level: The method used to construct this interval estimate is successful in capturing the actual value of the population proportion 90% of the time.

(c) The interval in part (b) is narrower because $\hat{p}$ of 0.76 is farther from 0.5 than $\hat{p}$ of 0.66. For a fixed sample size and confidence level, the confidence interval is widest when $\hat{p} = 0.5$, and decreases symmetrically as $\hat{p}$ moves closer to 0 or 1.

Chapter 10 Asking and Answering Questions about a Population Proportion

Section 10.1 Exercise Set 1

10.1: Hypotheses must be stated in terms of a population characteristic. This statement is incorrect because it is written in terms of a sample proportion ($\hat{p}$), and not a population characteristic (p).

10.2: $H_0 : p = \frac{1}{3}$ versus $H_a : p > \frac{1}{3}$, where p is the proportion of employers who had ever sent an employee home because they were dressed inappropriately.

10.3: $H_0 : p = 0.67$ versus $H_a : p \neq 0.67$, where p is the proportion of students at the high school who have a Facebook page that they update at least once a week.

10.4: (a) There is strong evidence that the proportion of American adults who favor drafting women if a military draft were reinstated is less than 0.5.

(b) Yes, it is reasonable to say that the data provide strong support for the alternative hypothesis.

(c) Yes, it is reasonable to say that the data provide strong evidence against the null hypothesis.

10.5: (a) The conclusion is consistent with testing

H_o: concealed weapons laws do not reduce crime

versus

H_a: concealed weapons laws reduce crime.

The null hypothesis is considered a statement of no difference, or equality. In this case, the null hypothesis is that there is no difference between states with and without concealed weapons laws (*i.e.*, the crime rates are equal). The research question asked if states that allow private citizens to carry concealed weapons result in a reduced crime rate, which is the alternative hypothesis.

(b) Recall that a statistical hypothesis test is capable of demonstrating strong support for the alternative hypothesis (by rejecting the null hypothesis). The statement "The strongest thing I could say is that I don't see any strong evidence that they are reducing crime" states that there is no strong evidence, so the null hypothesis was not rejected.

10.6: Rejecting the null hypothesis means that the sample data provide convincing evidence against the null hypothesis and in favor of the alternative hypothesis.

Section 10.1 Additional Exercises

10.13: (a) legitimate

(b) not legitimate

(c) not legitimate

(d) legitimate

(e) not legitimate

10.15: The pair of hypotheses that should be tested are $H_0: p = 0.6$ versus $H_a: p > 0.6$ because the research question indicates a desire for more than 60% of faculty favoring the change.

Section 10.2 Exercise Set 1

10.17: The other type of error is failing to reject the null hypothesis when it is false.

10.18: If the consequences of a Type I error are very serious, one would want to reduce the likelihood of making a Type I error. Since the probability of making a Type I error is α, we want to make α small.

10.19: (a) The alternative hypothesis $H_a: p > 0.1$ was chosen because the consumer group is carrying out the hypothesis test so see if there is strong evidence in favor of the proportion of defective flares being greater than 0.1.

(b)

Error	Definition of Error	Description of Error in context	Consequence of Error
Type I error	Reject a true H_o	Conclude that the proportion of defective flares is greater than 0.1 when it really is 0.1 or less.	The consumer group incorrectly files charges of false advertising against the manufacturer.

Type II error	Fail to reject a false H_o	You are not convinced that the proportion of defective flares is greater than 0.1 when it really is greater than 0.1.	The consumer group does not file charges of false advertising against the manufacturer when they should have.

10.20: (a) Type I error: The null hypothesis of $H_0: p = 0.9$ is incorrectly rejected in favor of the alternative hypothesis $H_a: p < 0.9$. The consumer group will claim false advertising even though the company's claim is actually true. Type II error: The null hypothesis of $H_0: p = 0.9$ is not rejected when it should have been. The consumer group will not claim false advertising even though less than 90% of all sets made by the manufacturer will not need service in the first 3 years.

(b) Since the consumer group does not want to claim false advertising unless there is strong evidence that $p < 0.9$, a test procedure with $\alpha = 0.01$ should be used. A smaller α value requires stronger evidence to reject the null hypothesis.

10.21: (a) This is a Type I error because the statement describes the result of incorrectly rejecting the null hypothesis. This probability is approximately

$$P(\text{Type I error}) \approx \frac{3}{33} = 0.091.$$

(b) The other error that is possible is a Type II error, in which the null hypothesis is not rejected when it should be. In this scenario, a Type II error is thinking that a woman has cancer in the other breast when she really does not have cancer in the other breast. This probability is approximately

$$P(\text{Type II error}) \approx \frac{91}{936} = 0.097$$

Section 10.2 Additional Exercises

10.27: A Type I error is rejecting H_o when H_o is true. A Type II error is failing to reject H_o when H_o is false.

10.29: Answers may vary. If the consequences of a Type II error are serious, then one should consider selecting a larger value for α. For example, a medication has serious side effects and the proportion of people on this medication who experience side effects is 0.07. A researcher thinks that if the medication is taken with food, the proportion experiencing side effects would be lower. They test

$H_0: p = 0.07$ versus $H_a: p < 0.07$. Since side effects are serious, you don't want to fail to reject the null if the proportion experiencing side effects is reduced.

10.31: It is not necessary to carry out a hypothesis test to determine if the proportion of felons released in 2003 was less than 0.40 because the 38% stated in the article is the proportion of all felons, which is a population proportion. This tells us that the population proportion is less than 0.4, so there is no need to carry out a hypothesis test.

Section 10.3 Exercise Set 1

10.32: (a) Shape: The shape of the sampling distribution is approximately normal because $np = 1,200(0.25) = 300 \geq 10$ and $n(1-p) = 1,200(1-0.25) = 900 \geq 10$.

Center: The sampling distribution is centered at $\mu_{\hat{p}} = 0.25$ because $\mu_{\hat{p}} = p$.

Spread: The standard deviation of the sampling distribution is

$$\sigma_{\hat{p}} = \sqrt{\frac{p(1-p)}{n}} = \sqrt{\frac{0.25(1-0.25)}{1200}} = 0.0125.$$

(b) I would not be surprised to observe a sample proportion of $\hat{p} = 0.24$ for a sample of size 1,200 if the null hypothesis $H_0: p = 0.25$ is true. First, compute a z statistic: $z = \dfrac{\hat{p}-p}{\sqrt{\dfrac{p(1-p)}{n}}} = \dfrac{0.24-0.25}{\sqrt{\dfrac{0.25(1-0.25)}{1,200}}} = -0.8$ (note that this z statistic is not particularly large in magnitude, and indicates that $\hat{p} = 0.24$ is less than one standard deviation below what you would expect it to be if the null hypothesis were true). Second, find $P(z \leq -0.8 \text{ when } H_0 \text{ is true}) = 0.212$. This probability is not small, indicating that it is not unlikely that we would observe a $\hat{p}$ at least as extreme as what we observed if H_0 is true.

(c) I would be surprised to observe a sample proportion of $\hat{p} = 0.20$ for a sample of size 1,200 if the null hypothesis $H_0: p = 0.25$ is true. First, compute a z statistic: $z = \dfrac{\hat{p}-p}{\sqrt{\dfrac{p(1-p)}{n}}} = \dfrac{0.20-0.25}{\sqrt{\dfrac{0.25(1-0.25)}{1,200}}} = -4$ (note that this z statistic is large in magnitude, four standard deviations below what you would expect it to be if the null hypothesis were true). Second, find $P(z \le -4 \text{ when } H_0 \text{ is true}) = 0.000032$. This probability indicates that it is highly unlikely we would observe a $\hat{p}$ at least as extreme as what we observed if H_0 is true.

(d) First, compute a z statistic: $z = \dfrac{\hat{p}-p}{\sqrt{\dfrac{p(1-p)}{n}}} = \dfrac{0.22-0.25}{\sqrt{\dfrac{0.25(1-0.25)}{1,200}}} = -2.4$ (note that this z statistic is 2.4 standard deviations below what you would expect it to be if the null hypothesis were true). Second, find $P(z \le -2.4 \text{ when } H_0 \text{ is true}) = 0.0082$. This probability indicates that it is unlikely we would observe a $\hat{p}$ at least as extreme as what we observed if H_0 is true. Because the computed probability is so small, there is convincing evidence that the goal is not being met.

10.33: (a) Shape: The sampling distribution is approximately normal because $np = 286(0.75) = 214.5 \ge 10$ and $n(1-p) = 286(1-0.75) = 71.5 \ge 10$.

Center: The sampling distribution is centered at $\mu_{\hat{p}} = 0.75$ because $\mu_{\hat{p}} = p$.

Spread: The standard deviation of the sampling distribution is $\sigma_{\hat{p}} = \sqrt{\dfrac{p(1-p)}{n}} = \sqrt{\dfrac{0.75(1-0.75)}{286}} = 0.0256$.

(b) I would be surprised to observe a sample proportion of $\hat{p} = 0.83$ for a sample of size 286 if the null hypothesis $H_0: p = 0.75$ is true. First, compute a z statistic: $z = \dfrac{\hat{p}-p}{\sqrt{\dfrac{p(1-p)}{n}}} = \dfrac{0.83-0.75}{\sqrt{\dfrac{0.75(1-0.75)}{286}}} = 3.124$ (note that this z statistic is large in magnitude, and indicates that $\hat{p} = 0.83$ is more than three standard deviations above what you would expect it to be if the null hypothesis were true). Second, find $P(z \ge 3.124 \text{ when } H_0 \text{ is true}) = 0.000892$. This probability indicates that it is highly unlikely we would observe a $\hat{p}$ at least as extreme as what we observed if H_0 is true.

(c) I would not be surprised to observe a sample proportion of $\hat{p} = 0.79$ for a sample of size 286 if the null hypothesis $H_0: p = 0.75$ is true. First, compute a z

statistic: $z=\frac{\hat{p}-p}{\sqrt{\frac{p(1-p)}{n}}}=\frac{0.79-0.75}{\sqrt{\frac{0.75(1-0.75)}{286}}}=1.56$ (note that this z statistic indicates that $\hat{p}=0.79$ is just over 1.5 standard deviations above what you would expect it to be if the null hypothesis were true). Second, find $P(z\geq 1.56 \text{ when } H_0 \text{ is true})=0.059$. This probability indicates that it is not particularly unlikely that we would observe a $\hat{p}$ at least as extreme as what we observed if H_0 is true.

(d) First, compute a z statistic: $z=\frac{\hat{p}-p}{\sqrt{\frac{p(1-p)}{n}}}=\frac{0.80-0.75}{\sqrt{\frac{0.75(1-0.75)}{286}}}=1.95$ (note that this z statistic indicates that $\hat{p}=0.80$ is just under 2 standard deviations above what you would expect it to be if the null hypothesis were true). Second, find $P(z\geq 1.95 \text{ when } H_0 \text{ is true})=0.026$. Because this computed probability is small, there is convincing evidence that the null hypothesis $H_0: p=0.75$ is not true.

Section 10.3 Additional Exercises

10.37: (a) Shape: The shape of the sampling distribution is approximately normal because $np=20(0.5)=10\geq 10$ and $n(1-p)=20(1-0.5)=10\geq 10$.

Center: The sampling distribution is centered at $\mu_{\hat{p}}=0.5$ because $\mu_{\hat{p}}=p$.

Spread: The standard deviation of the sampling distribution is $\sigma_{\hat{p}}=\sqrt{\frac{p(1-p)}{n}}=\sqrt{\frac{0.5(1-0.5)}{20}}=0.1118$.

(b) Answers will vary. One reasonable answer would be sample proportions greater than 0.7236 (these are sample proportions that are more than two standard deviations above 0.5).

Section 10.4 Exercise Set 1

10.38: (a) A P-value of 0.0003 means that it is very unlikely (probability = 0.0003), assuming that H_0 is true, that you would get a sample result at least as inconsistent with H_0 as the one obtained in the study. H_0 would be rejected.

(b) A P-value of 0.350 means that it is not particularly unlikely (probability = 0.350), assuming that H_0 is true, that you would get a sample result at least as inconsistent with H_0 as the one obtained in the study. There is no reason to reject H_0.

10.39: (a) $H_0 : p = \frac{2}{3}$ versus $H_a : p > \frac{2}{3}$, where p represents the proportion of American adults who oppose reinstatement of the draft.

(b) Because the P-value of 0.013 is less than α of 0.05, reject H_o. There is sufficient evidence to conclude that the proportion of American adults who oppose reinstatement of the draft is greater than $\frac{2}{3}$.

10.40: Completing Step 2 requires answering the four key questions (QSTN). (1) Question Type (Q) – is this an estimation or hypothesis testing question; (2) Study Type (S) – does the question use sample data or experiment data; (3) Type of Data (T) – is there one variable or two, and is the variable categorical or numerical; (4) Number of Samples/Treatments (N) – how many samples or treatments?

Section 10.4 Additional Exercises

10.45: A P-value of 0.0002 is strong evidence against the null hypothesis because the P-value is the probability, computed assuming the null hypothesis is true, of obtaining a test statistic value at least as extreme as what actually resulted. The smaller this probability, the less likely it is that we would observe a test statistic at least as extreme as what we did observe assuming the null hypothesis was true, and therefore the stronger the evidence is against the null hypothesis.

Section 10.5 Exercise Set 1

10.47: Question Type (Q): Estimation

Study Type (S): Sample data

Type of Data (T): One numerical variable

Number of Samples or Treatments (N): One sample

A hypothesis test for a population proportion would not be appropriate.

10.48: Question Type (Q): Hypothesis testing

Study Type (S): Sample data

Type of Data (T): One categorical variable

Number of Samples or Treatments (N): One sample

A hypothesis test for a population proportion would be appropriate.

10.49: Question Type (Q): Estimation

Study Type (S): Sample data

Type of Data (T): One categorical variable

Number of Samples or Treatments (N): One sample

A hypothesis test for a population proportion would not be appropriate.

10.50: (a) Large-sample z test is not appropriate because $np = 25(0.2) = 5$, which is less than 10.

(b) Large-sample z test is appropriate because $np = 200(0.6) = 120$ and $n(1-p) = 200(1-0.6) = 80$ are both at least 10.

(c) Large-sample z test is appropriate because $np = 100(0.9) = 90$ and $n(1-p) = 100(1-0.9) = 10$ are both at least 10.

(d) Large-sample z test is not appropriate because $np = 75(0.05) = 3.75$, which is less than 10.

10.51: (a) $P(z \le -0.55) = 0.2912$

(b) $P(z \le -0.92) = 0.1788$

(c) $P(z \le -1.99) = 0.0233$

(d) $P(z \le -2.24) = 0.0125$

(e) $P(z \le 1.40) = 0.9192$

10.52: (a) Using the 5-step process (HMC^3):

Hypotheses (H): We want to determine if there is evidence that the proportion of adult Americans who can answer the question correctly is less than 0.40. Therefore, the population characteristic of interest, *p*, is the proportion of adult Americans who can answer the question correctly. Test the hypotheses $H_0: p = 0.40$ versus $H_a: p < 0.40$.

Method (M): Because the answers to the four key questions are (Q) hypothesis testing, (S) sample data, (T) one categorical variable, and (N) one sample, we can consider a large-sample test for a population proportion. The test statistic is: $z = \frac{\hat{p} - 0.4}{\sqrt{\frac{0.4(1-0.4)}{n}}}$, and we will use a significance level of $\alpha = 0.05$.

Check (C): We are told that it is reasonable to regard the sample as representative of adult Americans. In addition, $np = 1{,}000(0.4) = 400$ and $n(1-p) = 1{,}000(1-0.4) = 600$, which are both at least 10. The two required conditions are satisfied.

Calculations (C): The sample proportion is $\hat{p} = \frac{354}{1{,}000} = 0.354$ and $n = 1{,}000$, so the test statistic is $z = \frac{0.354 - 0.4}{\sqrt{\frac{0.4(1-0.4)}{1{,}000}}} = -2.969$. This is a lower-tailed test (the inequality in H_a is <), so the *P*-value is the area under the *z* curve and to the left of –2.969. Therefore, the *P*-value is $P(z \le -2.969) = 0.0015$.

Communicate Results (C): Because the *P*-value of 0.0015 is less than the selected significance level of $\alpha = 0.05$, we reject the null hypothesis. There is convincing evidence that the proportion of adult Americans who can answer the question correctly is less than 0.4.

(b) Using the 5-step process (HMC^3):

Hypotheses (H): We want to determine if there is evidence that the proportion of adult Americans who would choose one of the wrong answers (a, b, or c) is more than 1/3. Therefore, the population characteristic of interest, *p*, is the proportion of adult Americans would choose one of the wrong answers (a, b, or c). Test the hypotheses $H_0 : p = \frac{1}{3}$ versus $H_a : p > \frac{1}{3}$.

Method (M): Because the answers to the four key questions are (Q) hypothesis testing, (S) sample data, (T) one categorical variable, and (N) one sample, we can consider a large-sample test for a population proportion. The test statistic is:
$z = \dfrac{\hat{p} - (1/3)}{\sqrt{\dfrac{(1/3)(1-(1/3))}{n}}}$, and we will use a significance level of $\alpha = 0.05$.

Check (C): We are told that it is reasonable to regard the sample as representative of adult Americans. In addition, the sample proportion is $\hat{p} = 0.378$, so $np = 1{,}000(1/3) = 333.333$ and $n(1-p) = 1{,}000(1-(1/3)) = 666.667$, which are both at least 10. The two required conditions are satisfied.

Calculations (C): The sample proportion is $\hat{p} = 0.378$ and $n = 1{,}000$, so the test statistic is $z = \dfrac{0.378 - (1/3)}{\sqrt{\dfrac{(1/3)(1-(1/3))}{1{,}000}}} = 2.996$. This is an upper-tailed test (the inequality in H_a is >), so the P-value is the area under the z curve and to the right of 2.996. Therefore, the P-value is $P(z \geq 2.996) = 0.0014$.

Communicate Results (C): Because the P-value of 0.0014 is less than the selected significance level of $\alpha = 0.05$, we reject the null hypothesis. There is convincing evidence that the proportion of adult Americans who would choose one of the wrong answers (a, b, or c) is more than 1/3.

10.53: Using the 5-step process (HMC^3):

Hypotheses (H): We want to determine if there is evidence that the proportion of 16- to 17-year-old Americans that sent a text message while driving is more than 1/4. Therefore, the population characteristic of interest, p, is the proportion of 16 to 17 year old teens that sent a text message while driving. Test the hypotheses $H_0 : p = 0.25$ versus $H_a : p > 0.25$.

Method (M): Because the answers to the four key questions are (Q) hypothesis testing, (S) sample data, (T) one categorical variable, and (N) one sample, we can consider a large-sample test for a population proportion. The test statistic is:

$$z = \frac{\hat{p} - 0.25}{\sqrt{\frac{0.25(1-0.25)}{n}}}$$, and we will use a significance level of $\alpha = 0.01$.

Check (C): We are told to assume that the sample is a simple random sample of 16- to 17-year-old Americans. In addition, $np = 283(0.25) = 70.75$ and $n(1-p) = 283(1-0.25) = 212.25$, which are both at least 10. The two required conditions are satisfied.

Calculations (C): The sample proportion is $\hat{p} = \frac{74}{283} = 0.2615$ and $n = 283$, so the test statistic is $z = \frac{0.2615 - 0.25}{\sqrt{\frac{0.25(1-0.25)}{283}}} = 0.447$. This is an upper-tailed test (the inequality in H_a is >), so the P-value is the area under the z curve and to the right of 0.447. Therefore, the P-value is $P(z \geq 0.447) = 0.327$.

Communicate Results (C): Because the P-value of 0.327 is greater than the selected significance level of $\alpha = 0.01$, we fail to reject the null hypothesis. There is not convincing evidence that the proportion of 16- to 17-year-old Americans that sent a text message while driving is more than 0.25.

10.54: Using the 5-step process (HMC^3):

Hypotheses (H): We want to determine if there is evidence that a majority of adult Americans prefer to watch movies at home rather than at a movie theater. Therefore, the population characteristic of interest, *p*, is the proportion of adult Americans who prefer to watch movies at home rather than at a movie theater. Test the hypotheses $H_0 : p = 0.5$ versus $H_a : p > 0.5$.

Method (M): Because the answers to the four key questions are (Q) hypothesis testing, (S) sample data, (T) one categorical variable, and (N) one sample, we can consider a large-sample test for a population proportion. The test statistic is:

$$z = \frac{\hat{p} - 0.5}{\sqrt{\frac{0.5(1-0.5)}{n}}}$$, and we will use a significance level of $\alpha = 0.05$.

Check (C): We are told that the sample consisted of 1,000 randomly selected adult Americans. In addition, $np = 1,000(0.5) = 500$ and $n(1-p) = 1,000(1-0.5) = 500$, which are both at least 10. The two required conditions are satisfied.

Calculations (C): The sample proportion is $\hat{p} = \frac{730}{1000} = 0.730$ and $n = 1,000$, so the test statistic is $z = \frac{0.730 - 0.5}{\sqrt{\frac{0.5(1-0.5)}{1,000}}} = 14.546$. This is an upper-tailed test (the inequality in H_a is >), so the P-value is the area under the z curve and to the right of 14.546. Therefore, the P-value is $P(z \geq 14.546) \approx 0$.

Communicate Results (C): Because the P-value of approximately 0 is less than the selected significance level of $\alpha = 0.05$, we reject the null hypothesis. There is convincing evidence that a majority of adult Americans prefer to watch movies at home rather than at a movie theater.

10.55: (a) Using the 5-step process (HMC^3):

Hypotheses (H): We want to determine if there is evidence that less than half of adult Americans believe that movie quality is getting worse. Therefore, the population characteristic of interest, p, is the proportion of adult Americans believe that movie quality is getting worse. Test the hypotheses $H_0: p = 0.5$ versus $H_a: p < 0.5$.

Method (M): Because the answers to the four key questions are (Q) hypothesis testing, (S) sample data, (T) one categorical variable, and (N) one sample, we can consider a large-sample test for a population proportion. The test statistic is: $z = \frac{\hat{p} - 0.5}{\sqrt{\frac{0.5(1-0.5)}{n}}}$, and we will use a significance level of $\alpha = 0.05$.

Check (C): We are told that the sample consisted of 1,000 randomly selected adult Americans. In addition, $np = 1,000(0.5) = 500$ and $n(1-p) = 1,000(1-0.5) = 500$, which are both at least 10. The two required conditions are satisfied.

Calculations (C): The sample proportion is $\hat{p} = \dfrac{470}{1000} = 0.470$ and $n = 1{,}000$, so the test statistic is $z = \dfrac{0.470 - 0.5}{\sqrt{\dfrac{0.5(1-0.5)}{1{,}000}}} = -1.897$. This is a lower-tailed test (the inequality in H_a is <), so the P-value is the area under the z curve and to the left of -1.897. Therefore, the P-value is $P(z \le -1.897) = 0.0289$.

Communicate Results (C): Because the P-value of 0.0289 is less than the selected significance level of $\alpha = 0.05$, we reject the null hypothesis. There is convincing evidence that less than half of adult Americans believe that movie quality is getting worse.

(b) Using the 5-step process (HMC^3):

Hypotheses (H): We want to determine if there is evidence that less than half of adult Americans believe that movie quality is getting worse. Therefore, the population characteristic of interest, p, is the proportion of adult Americans believe that movie quality is getting worse. Test the hypotheses $H_0 : p = 0.5$ versus $H_a : p < 0.5$.

Method (M): Because the answers to the four key questions are (Q) hypothesis testing, (S) sample data, (T) one categorical variable, and (N) one sample, we can consider a large-sample test for a population proportion. The test statistic is: $z = \dfrac{\hat{p} - 0.5}{\sqrt{\dfrac{0.5(1-0.5)}{n}}}$, and we will use a significance level of $\alpha = 0.05$.

Check (C): We are told to assume that the sample consists of 100 randomly selected adult Americans. In addition, $np = 100(0.5) = 50$ and $n(1-p) = 100(1-0.5) = 50$, which are both at least 10. The two required conditions are satisfied.

Calculations (C): The sample proportion is $\hat{p} = \dfrac{47}{100} = 0.47$ and $n = 100$, so the test statistic is $z = \dfrac{0.47 - 0.5}{\sqrt{\dfrac{0.5(1-0.5)}{100}}} = -0.6$. This is a lower-tailed test (the inequality in H_a is <), so the P-value is the area under the z curve and to the left of -0.6. Therefore, the P-value is $P(z \le -0.6) = 0.274$.

Communicate Results (C): Because the P-value of 0.274 is greater than the selected significance level of $\alpha = 0.05$, we fail to reject the null hypothesis. There is not convincing evidence that less than half of adult Americans believe that movie quality is getting worse.

(c) Both results *suggest* that fewer than half of adult Americans believe that movie quality is getting worse. However, getting 470 out of 1000 people responding this way (as opposed to 47 out of 100) provides much *stronger* evidence of this fact.

10.56: Using the 5-step process (HMC^3):

Hypotheses (H): We want to determine if there is evidence that less than half of adult Americans are somewhat or very interested in having Web access in their cars. Therefore, the population characteristic of interest, p, is the proportion of adult Americans who are somewhat or very interested in having Web access in their cars. Test the hypotheses $H_0 : p = 0.5$ versus $H_a : p < 0.5$.

Method (M): Because the answers to the four key questions are (Q) hypothesis testing, (S) sample data, (T) one categorical variable, and (N) one sample, we can consider a large-sample test for a population proportion. The test statistic is: $z = \dfrac{\hat{p} - 0.5}{\sqrt{\dfrac{0.5(1-0.5)}{n}}}$, and we will use a significance level of $\alpha = 0.05$.

Check (C): We are told to assume that the sample is representative of adult Americans. In addition, $np = 1{,}005(0.5) = 502.5$ and $n(1-p) = 1{,}005(1-0.5) = 502.5$, which are both at least 10. The two required conditions are satisfied.

Calculations (C): The sample proportion is $\hat{p} = 0.46$ and $n = 1{,}005$, so the test statistic is $z = \dfrac{0.46 - 0.5}{\sqrt{\dfrac{0.5(1-0.5)}{1{,}005}}} = -2.536$. This is a lower-tailed test (the inequality in H_a is <), so the P-value is the area under the z curve and to the left of –2.536. Therefore, the P-value is $P(z \le -2.536) = 0.0056$.

Communicate Results (C): Because the P-value of 0.0056 is less than the selected significance level of $\alpha = 0.05$, we reject the null hypothesis. There is convincing evidence that less than half of adult Americans are somewhat or very interested in having Web access in their cars. The marketing manager is not correct in his claim.

Section 10.5 Additional Exercises

10.67: (a) Using the 5-step process (HMC^3):

Hypotheses (H): We want to determine if there is evidence that a majority of women ages 22 to 35 who work full time would be willing to give up some personal time for more money. Therefore, the population characteristic of interest, p, is the proportion of women ages 22 to 35 who work full time who would be willing to give up some personal time for more money. Test the hypotheses $H_0 : p = 0.5$ versus $H_a : p > 0.5$.

Method (M): Because the answers to the four key questions are (Q) hypothesis testing, (S) sample data, (T) one categorical variable, and (N) one sample, we can consider a large-sample test for a population proportion. The test statistic is: $z = \dfrac{\hat{p} - 0.5}{\sqrt{\dfrac{0.5(1-0.5)}{n}}}$, and we will use a significance level of $\alpha = 0.01$.

Check (C): We are told that the sample of size 1,000 was selected to be representative of women ages 22 to 35 who work full time. In addition, $np = 1{,}000(0.5) = 500$ and $n(1-p) = 1{,}000(1-0.5) = 500$, which are both at least 10. The two required conditions are satisfied.

Calculations (C): The sample proportion is $\hat{p} = \dfrac{540}{1000} = 0.540$ and $n = 1{,}000$, so the test statistic is $z = \dfrac{0.54 - 0.5}{\sqrt{\dfrac{0.5(1-0.5)}{1{,}000}}} = 2.530$. This is an upper-tailed test (the inequality in H_a is >), so the P-value is the area under the z curve and to the right of 2.530. Therefore, the P-value is $P(z \geq 2.530) = 0.0057$.

Communicate Results (C): Because the P-value of 0.0057 is less than the selected significance level of $\alpha = 0.01$, we reject the null hypothesis. There is convincing evidence that a majority of women ages 22 to 35 who work full time would be willing to give up some personal time for more money.

(b) It is not reasonable to generalize to the population of all working women because the sample was selected to be representative of women ages 22 to 35 who work full time. To generalize to the population of all working women, the sample would need to be representative of all working women.

10.69: Using the 5-step process (HMC^3):

Hypotheses (H): We want to determine if there is evidence that more than three-quarters of American adults believe that lack of respect and courtesy is a serious problem. Therefore, the population characteristic of interest, *p*, is the proportion American adults who believe that lack of respect and courtesy is a serious problem. Test the hypotheses $H_0: p = 0.75$ versus $H_a: p > 0.75$.

Method (M): Because the answers to the four key questions are (Q) hypothesis testing, (S) sample data, (T) one categorical variable, and (N) one sample, we can consider a large-sample test for a population proportion. The test statistic is:
$z = \dfrac{\hat{p} - 0.75}{\sqrt{\dfrac{0.75(1-0.75)}{n}}}$, and we will use a significance level of $\alpha = 0.05$.

Check (C): We are told that the sample of size 2,013 was selected to be representative of American adults. In addition, $np = 2{,}013(0.75) = 1509.75$ and $n(1-p) = 2{,}013(1-0.75) = 503.25$, which are both at least 10. The two required conditions are satisfied.

Calculations (C): The sample proportion is $\hat{p} = \dfrac{1{,}590}{2{,}013} = 0.790$ and $n = 2{,}013$, so the test statistic is $z = \dfrac{0.790 - 0.75}{\sqrt{\dfrac{0.75(1-0.75)}{2{,}013}}} = 4.145$. This is an upper-tailed test (the inequality in H_a is >), so the *P*-value is the area under the *z* curve and to the right of 4.145. Therefore, the *P*-value is $P(z > 4.145) = 0.00002$.

Communicate Results (C): Because the *P*-value of 0.00002 is less than the selected significance level of $\alpha = 0.05$, we reject the null hypothesis. There is convincing evidence that more than three-quarters of American adults believe that lack of respect and courtesy is a serious problem.

10.71: Using the 5-step process (HMC^3):

Hypotheses (H): We want to determine if there is evidence that more than two-thirds of American adults approve of casino gambling. Therefore, the population characteristic of interest, p, is the proportion American adults who approve of casino gambling. Test the hypotheses $H_0: p = \frac{2}{3}$ versus $H_a: p > \frac{2}{3}$.

Method (M): Because the answers to the four key questions are (Q) hypothesis testing, (S) sample data, (T) one categorical variable, and (N) one sample, we can consider a large-sample test for a population proportion. The test statistic is: $z = \frac{\hat{p} - (2/3)}{\sqrt{\frac{(2/3)(1-(2/3))}{n}}}$, and we will use a significance level of $\alpha = 0.05$.

Check (C): We are told that the sample of size 1,523 was randomly selected from the population of American adults. In addition, $np = 1,523\left(\frac{2}{3}\right) = 1015.33$ and $n(1-p) = 1,523\left(1 - \frac{2}{3}\right) = 507.67$, which are both at least 10. The two required conditions are satisfied.

Calculations (C): The sample proportion is $\hat{p} = \frac{1,035}{1,523} = 0.6796$ and $n = 1,523$, so the test statistic is $z = \frac{0.6796 - (2/3)}{\sqrt{\frac{(2/3)(1-(2/3))}{1,523}}} = 1.0707$. This is an upper-tailed test (the inequality in H_a is >), so the *P*-value is the area under the *z* curve and to the right of 1.0707. Therefore, the *P*-value is $P(z \geq 1.0707) = 0.142$.

Communicate Results (C): Because the *P*-value of 0.142 is greater than the selected significance level of $\alpha = 0.05$, we fail to reject the null hypothesis. There is not convincing evidence that more than two-thirds of American adults approve of casino gambling.

10.73: Using the 5-step process (HMC^3):

Hypotheses (H): We want to determine if there is evidence that more than one-fourth of American adults see a lottery or sweepstakes as their best chance of accumulating \$500,000. Therefore, the population characteristic of interest, *p*, is the proportion American adults who see a lottery or sweepstakes as their best chance of accumulating \$500,000. Test the hypotheses $H_0: p = 0.25$ versus $H_a: p > 0.25$.

Method (M): Because the answers to the four key questions are (Q) hypothesis testing, (S) sample data, (T) one categorical variable, and (N) one sample, we can consider a large-sample test for a population proportion. The test statistic is: $z = \dfrac{\hat{p} - 0.25}{\sqrt{\dfrac{0.25(1-0.25)}{n}}}$, and we will use a significance level of $\alpha = 0.01$.

Check (C): We are told that the sample of size 1,010 was randomly selected from the population of American adults. In addition, $np = 1{,}010(0.25) = 252.5$ and $n(1-p) = 1{,}010(1-0.25) = 757.5$, which are both at least 10. The two required conditions are satisfied.

Calculations (C): The sample proportion is $\hat{p} = 0.28$ and $n = 1{,}010$, so the test statistic is $z = \dfrac{0.28 - 0.25}{\sqrt{\dfrac{0.25(1-0.25)}{1{,}010}}} = 2.202$. This is an upper-tailed test (the inequality in H_a is >), so the *P*-value is the area under the *z* curve and to the right of 2.202. Therefore, the *P*-value is $P(z \geq 2.202) = 0.0138$.

Communicate Results (C): Because the *P*-value of 0.0138 is greater than the selected significance level of $\alpha = 0.01$, we fail to reject the null hypothesis. There is not convincing evidence that more than one-fourth of American adults see a lottery or sweepstakes as their best chance of accumulating \$500,000.

10.75: Using the 5-step process (HMC^3):

Hypotheses (H): We want to determine if there is evidence that the proportion of religion surfers who belong to a religious community is different from 0.68. Therefore, the population characteristic of interest, *p*, is the proportion of religion surfers who belong to a religious community. Test the hypotheses $H_0: p = 0.68$ versus $H_a: p \neq 0.68$.

Method (M): Because the answers to the four key questions are (Q) hypothesis testing, (S) sample data, (T) one categorical variable, and (N) one sample, we can consider a large-sample test for a population proportion. The test statistic is:
$z = \frac{\hat{p} - 0.68}{\sqrt{\frac{0.68(1-0.68)}{n}}}$, and we will use a significance level of $\alpha = 0.05$.

Check (C): We are told that the sample of size 512 is representative of religion surfers. In addition, $np = 512(0.68) = 348.16$ and $n(1-p) = 512(1-0.68) = 163.84$, which are both at least 10. The two required conditions are satisfied.

Calculations (C): The sample proportion is $\hat{p} = 0.84$ and $n = 512$, so the test statistic is $z = \frac{0.84 - 0.68}{\sqrt{\frac{0.68(1-0.68)}{512}}} = 7.761$. This is a two-tailed test (the inequality in H_a is $\neq$), so the *P*-value is the total area under the *z* curve and to the left of –7.761 and to the right of 7.761. Therefore, by the symmetry of the normal curve, the *P*-value is $2 \cdot P(z \leq -7.761) \approx 0$.

Communicate Results (C): Because the *P*-value of approximately 0 is less than the selected significance level of $\alpha = 0.05$, we reject the null hypothesis. There is convincing evidence that the proportion of religion surfers who belong to a religious community is different from 0.68.

Are You Ready to Move On? Chapter 10 Review Exercises

10.77: $H_0: p = 0.37$ versus $H_a: p \neq 0.37$, where *p* represents the proportion of the local high school students who access the Internet from a mobile phone.

10.79: Failing to reject the null hypothesis means that you are not convinced that the null hypothesis is false. This is not the same as being convinced that it is true.

10.81: (a) The phrase "From the data at hand, there was no convincing evidence of any increased risk of death from any of the cancers surveyed due to living near nuclear facilities" indicates that the researchers failed to reject the null hypothesis.

(b) If the Cancer Institute researchers are incorrect in their conclusion, then they would have failed to reject H_0 when H_0 was actually false. This is a Type II error.

(c) Yes. The study did not provide sufficient evidence to reject the null hypothesis. In other words, there was not convincing evidence that the proportion of the population in areas near nuclear power plants who die of cancer during a given year is greater than the proportion of the population in areas where no nuclear power plants are present. This does not mean that there was no effect, only that the data did not provide evidence to reject the null hypothesis.

10.83: (a) Shape: The shape of the sampling distribution is approximately normal because $np = 728(0.25) = 182 \geq 10$ and $n(1-p) = 728(1-0.25) = 546 \geq 10$.

Center: The sampling distribution is centered at $\mu_{\hat{p}} = 0.25$ because $\mu_{\hat{p}} = p$.

Spread: The standard deviation of the sampling distribution is

$$\sigma_{\hat{p}} = \sqrt{\frac{p(1-p)}{n}} = \sqrt{\frac{0.25(1-0.25)}{728}} = 0.0160.$$

(b) I would not be surprised to observe a sample proportion of $\hat{p} = 0.27$ for a sample of size 728 if the null hypothesis $H_0 : p = 0.25$ is true. First, compute a z statistic: $z = \dfrac{\hat{p}-p}{\sqrt{\dfrac{p(1-p)}{n}}} = \dfrac{0.27-0.25}{\sqrt{\dfrac{0.25(1-0.25)}{728}}} = 1.2462$ (note that this z statistic indicates that $\hat{p} = 0.27$ is approximately 1.25 standard deviations above what you would expect it to be if the null hypothesis were true). Second, find $P(z \geq 1.2462 \text{ when } H_0 \text{ is true}) = 0.1063$. This probability indicates that it is not unlikely that we would observe a $\hat{p}$ at least as extreme as what we observed if H_0 is true.

(c) I would be surprised to observe a sample proportion of $\hat{p} = 0.31$ for a sample of size 728 if the null hypothesis $H_0 : p = 0.25$ is true. First, compute a z statistic: $z = \dfrac{\hat{p}-p}{\sqrt{\dfrac{p(1-p)}{n}}} = \dfrac{0.31-0.25}{\sqrt{\dfrac{0.25(1-0.25)}{728}}} = 3.74$ (note that this z statistic indicates that $\hat{p} = 0.31$ is approximately 3.74 standard deviations above what you would expect it to be if the null hypothesis were true). Second, find

$P(z \geq 3.74 \text{ when } H_0 \text{ is true}) = 0.000092$. This probability indicates that it is quite unlikely that we would observe a $\hat{p}$ at least as extreme as what we observed if H_0 is true.

(d) First, compute a z statistic: $z = \dfrac{\hat{p}-p}{\sqrt{\dfrac{p(1-p)}{n}}} = \dfrac{0.33-0.25}{\sqrt{\dfrac{0.25(1-0.25)}{728}}} = 4.985$ (note that this z statistic indicates that $\hat{p} = 0.33$ is just under 5 standard deviations above what you would expect it to be if the null hypothesis were true). Second, find $P(z \geq 4.985 \text{ when } H_0 \text{ is true}) \approx 0$. Because this computed probability is essentially zero, there is convincing evidence that more than 25% of law enforcement agencies review social media activity as part of background checks.

10.85: Question Type (Q): Hypothesis testing

Study Type (S): Sample data

Type of Data (T): One numerical variable

Number of Samples or Treatments (N): One sample

A hypothesis test for a population proportion would not be appropriate.

10.87: (a) $H_0: p = 0.25$ versus $H_a: p > 0.25$, where p represents the true proportion of college students who are "phantom smokers."

(b) Yes, because $np = 899(0.25) = 224.75$ and $n(1-p) = 899(1-0.25) = 674.25$, which are both greater than 10.

(c) The sample proportion is $\hat{p} = \dfrac{268}{899} = 0.2981$, so

$z = \dfrac{\hat{p}-p}{\sqrt{\dfrac{p(1-p)}{n}}} = \dfrac{0.2981-0.25}{\sqrt{\dfrac{0.25(1-0.25)}{899}}} = 3.33$, and the P-value is

$P(z \geq 3.33) = 0.0004 \approx 0$.

(d) Because the P-value of approximately 0 is less than α of 0.05, we reject H_0.

10.89: Using the 5-step process (HMC³):

Hypotheses (H): We want to determine if there is evidence that the response rate when the distributor wears an eye patch is greater than the previous rate of 40%. Therefore, the population characteristic of interest, *p*, is the true proportion of questionnaires returned when the distributor wears an eye patch. Test the hypotheses $H_0 : p = 0.40$ versus $H_a : p > 0.40$.

Method (M): Because the answers to the four key questions are (Q) hypothesis testing, (S) sample data, (T) one categorical variable, and (N) one sample, we can consider a large-sample test for a population proportion. The test statistic is: $z = \dfrac{\hat{p} - 0.40}{\sqrt{\dfrac{0.40(1-0.40)}{n}}}$, and we will use a significance level of $\alpha = 0.05$.

Check (C): We must assume that the sample of size 200 is representative of all questionnaire respondents. In addition, $np = 200(0.4) = 80$ and $n(1-p) = 200(1-0.4) = 120$, which are both at least 10. The two required conditions are satisfied.

Calculations (C): The sample proportion is $\hat{p} = \dfrac{109}{200} = 0.545$ and $n = 200$, so the test statistic is $z = \dfrac{0.545 - 0.40}{\sqrt{\dfrac{0.40(1-0.40)}{200}}} = 4.186$. This is an upper-tailed test (the inequality in H_a is >), so the *P*-value is the area under the *z* curve and to the right of 4.186. Therefore, the *P*-value is $P(z \geq 4.186) \approx 0$.

Communicate Results (C): Because the *P*-value of approximately 0 is less than the selected significance level of $\alpha = 0.05$, we reject the null hypothesis. There is convincing evidence that the response rate is greater than 40%.

Chapter 11 Asking and Answering Questions about the Difference between Two Population Proportions

Section 11.1 Exercise Set 1

11.1: (a) Question type (Q): Estimation

Study type (S): Sample data

Type of data (T): One categorical variable

Number of samples or treatments (N): Two samples (one from the population of men and from the population of women)

This particular combination of answers to QSTN suggests that a large-sample confidence interval for a difference of two population proportions is appropriate.

(b) Using the 5-step process (EMC^3):

Estimate (E): The difference in proportions living in poverty for men and women, $p_1 - p_2$, will be estimated, where p_1 is the true proportion of men living in poverty and p_2 is the true proportion of women living in poverty.

Method (M): Because the answers to the four key questions are (Q) estimation, (S) sample data, (T) one categorical variable, and (N) two samples, we can construct and interpret a 90% large-sample confidence interval for the difference in proportions living in poverty for men and women.

Check (C): We are told that the data are from large representative samples of men and women. In addition, $n_1\hat{p}_1 = 1,200(0.125) = 150$, $n_1(1-\hat{p}_1) = 1,200(1-0.125) = 1,050$, $n_2 p_2 = 1,000(0.151) = 151$, and $n_2(1-\hat{p}_2) = 1,000(1-0.151) = 849$, which are all at least 10. The two required conditions are satisfied.

Calculations (C):

$$(\hat{p}_1 - \hat{p}_2) \pm (z \text{ critical value})\sqrt{\frac{\hat{p}_1(1-\hat{p}_1)}{n_1} + \frac{\hat{p}_2(1-\hat{p}_2)}{n_2}}$$

$$(0.125 - 0.151) \pm 1.645\sqrt{\frac{0.125(1-0.125)}{1,200} + \frac{0.151(1-0.151)}{1,000}}$$

$$-0.026 \pm 1.645(0.01481)$$

$$-0.026 \pm 0.024363$$

$$(-0.050, -0.002)$$

Communicate Results (C):

Interpret confidence interval: We are 90% confident that the actual difference in proportions living in poverty for men and women is somewhere between –0.0504 and –0.0016. Because both endpoints of the confidence interval are negative, we believe that the percent of women living in poverty is greater than the percent of men living in poverty by somewhere between 0.2% and 5.0%.

Interpret confidence level: The method used to construct this interval estimate is successful in capturing the actual value of the difference in population proportions 90% of the time.

11.2: (a) Yes. Let $\hat{p}_1 = 0.20$ and $\hat{p}_2 = 0.15$ and $n_1 = n_2 = 1,112$. Since $n_1\hat{p}_1 = 1,112(0.20) = 222.4$, $n_1(1-\hat{p}_1) = 1,112(1-0.20) = 889.6$, $n_2\hat{p}_2 = 1,112(0.15) = 166.8$, and $n_2(1-\hat{p}_2) = 1,112(1-0.15) = 945.2$ are all at least 10, the sample sizes are large enough to use the large-sample confidence interval.

(b) The difference in proportions of Americans ages 12 and older who owned MP3 player in 2006 and the corresponding proportion for 2005, $p_1 - p_2$, will be estimated, where p_1 is the true proportion of Americans ages 12 and older who owned MP3 player in 2006 and p_2 is the true proportion of Americans ages 12 and older who owned MP3 player in 2005. Let $\hat{p}_1 = 0.20$ and $\hat{p}_2 = 0.15$.

$$(\hat{p}_1 - \hat{p}_2) \pm (z \text{ critical value})\sqrt{\frac{\hat{p}_1(1-\hat{p}_1)}{n_1} + \frac{\hat{p}_2(1-\hat{p}_2)}{n_2}}$$

$$(0.20 - 0.15) \pm 1.96\sqrt{\frac{0.20(1-0.20)}{1{,}112} + \frac{0.15(1-0.15)}{1{,}112}}$$

$$0.05 \pm 1.96(0.016079)$$

$$0.05 \pm 0.031515$$

$$(0.018, 0.082)$$

(c) Zero is not included in the confidence interval. This means that you can be confident that the proportion of Americans ages 12 and older who owned an MP3 player was greater in 2006 than in 2005 by somewhere between 0.018 and 0.082.

(d) We are 95% confident that the actual difference in proportions of Americans ages 12 and older who owned MP3 player in 2006 and the corresponding proportion for 2005 is somewhere between 0.018 and 0.082. Because both endpoints of the confidence interval are positive, we believe that the percent of Americans ages 12 and older who owned MP3 player in 2006 is greater than the corresponding proportion in 2005 by somewhere between 1.8% and 8.2%.

11.3: (a) Using the 5-step process (EMC^3):

Estimate (E): The difference between the proportion of avid mountain bikers with low sperm count and the proportion for nonbikers, $p_1 - p_2$, will be estimated, where p_1 is the true proportion of avid mountain bikers with low sperm count and p_2 is the true proportion nonbikers with low sperm count.

Method (M): Because the answers to the four key questions are (Q) estimation, (S) sample data, (T) one categorical variable, and (N) two samples, we can construct and interpret a 95% large-sample confidence interval for the difference in proportions of avid mountain bikers with low sperm count and the proportion for nonbikers.

Check (C): We are told that the samples are representative of the populations of avid mountain bikers and nonbikers. In addition, $n_1\hat{p}_1 = 100(0.9) = 90$, $n_1(1-\hat{p}_1) = 100(1-0.9) = 10$, $n_2 p_2 = 100(0.26) = 26$, and $n_2(1-\hat{p}_2) = 100(1-0.26) = 74$, which are all at least 10. The two required conditions are satisfied.

Calculations (C):

$$(\hat{p}_1 - \hat{p}_2) \pm (z \text{ critical value})\sqrt{\frac{\hat{p}_1(1-\hat{p}_1)}{n_1} + \frac{\hat{p}_2(1-\hat{p}_2)}{n_2}}$$

$$(0.90 - 0.26) \pm 1.96\sqrt{\frac{0.90(1-0.90)}{100} + \frac{0.26(1-0.26)}{100}}$$

$$0.64 \pm 1.96(0.053141)$$

$$0.64 \pm 0.104157$$

$$(0.536, 0.744)$$

Communicate Results (C):

Interpret confidence interval: We are 95% confident that the actual difference in the proportion of avid bikers with low sperm count and nonbikers with low sperm count is somewhere between 0.536 and 0.744. Because both endpoints of the confidence interval are positive, we believe that the percent of avid bikers with low sperm count is greater than the percent of nonbikers with low sperm count by somewhere between 53.6% and 74.4%.

Interpret confidence level: The method used to construct this interval estimate is successful in capturing the actual value of the difference in population proportions 95% of the time.

(b) No, because this was an observational study and it is not a good idea to draw cause-and-effect conclusions from an observational study.

Section 11.1 Additional Exercises

11.7: Using the 5-step process (EMC^3):

Estimate (E): The difference between the proportion of high school students who believed marijuana was very distracting or extremely distracting in 2009 and this proportion in 2011, $p_1 - p_2$, will be estimated, where p_1 is the true proportion of high school students who believed marijuana was very distracting or extremely distracting in 2009 and p_2 is the true proportion of high school students who believed marijuana was very distracting or extremely distracting in 2011.

Method (M): Because the answers to the four key questions are (Q) estimation, (S) sample data, (T) one categorical variable, and (N) two samples, we can construct and interpret a 99% large-sample confidence interval for the difference in proportions of high school students who believed marijuana was very distracting or extremely distracting in 2009 and this proportion in 2011.

Check (C): We are told that the samples are representative of the populations high school students in 2009 and 2011. In addition, $n_1\hat{p}_1 = 2,300(0.78) = 1794$, $n_1(1-\hat{p}_1) = 2,300(1-0.78) = 506$, $n_2 p_2 = 2,294(0.70) = 1605.8$, and $n_2(1-\hat{p}_2) = 2,294(1-0.70) = 688.2$, which are all at least 10. The two required conditions are satisfied.

Calculations (C):

$$(\hat{p}_1 - \hat{p}_2) \pm (z \text{ critical value})\sqrt{\frac{\hat{p}_1(1-\hat{p}_1)}{n_1} + \frac{\hat{p}_2(1-\hat{p}_2)}{n_2}}$$

$$(0.78-0.70) \pm 2.58\sqrt{\frac{0.78(1-0.78)}{2,300} + \frac{0.70(1-0.70)}{2,294}}$$

$$0.08 \pm 2.58(0.01289)$$

$$0.08 \pm 0.033256$$

$$(0.047, 0.113)$$

Communicate Results (C):

Interpret confidence interval: We are 99% confident that the actual difference in the proportion of high school students who believed marijuana was very distracting or extremely distracting in 2009 and this proportion in 2011 is somewhere between 0.047 and 0.113. Because both endpoints of the confidence interval are positive, we believe that the percent of high school students who believed marijuana was very distracting or extremely distracting in 2009 is greater than the percent of high school students who believed marijuana was very distracting or extremely distracting in 2011 by somewhere between 4.7% and 11.3%.

Interpret confidence level: The method used to construct this interval estimate is successful in capturing the actual value of the difference in population proportions 99% of the time.

11.9: (a) Using the 5-step process (EMC^3):

Estimate (E): The difference between the proportion of high school graduates who were unemployed in 2008 and high school graduates who were unemployed in 2009, $p_1 - p_2$, will be estimated, where p_1 is the true proportion of high school graduates who were unemployed in 2008 and p_2 is the true proportion of high school graduates who were unemployed in 2009.

Method (M): Because the answers to the four key questions are (Q) estimation, (S) sample data, (T) one categorical variable, and (N) two samples, we can construct and interpret a 99% large-sample confidence interval for the difference between the proportion of high school graduates who were unemployed in 2008 and 2009.

Check (C): We are told that the samples are representative of the populations high school graduates in 2008 and 2009. In addition, $n_1\hat{p}_1 = 400(0.057) = 22.8$, $n_1(1-\hat{p}_1) = 400(1-0.057) = 377.2$, $n_2 p_2 = 400(0.097) = 38.8$, and $n_2(1-\hat{p}_2) = 400(1-0.097) = 361.2$, which are all at least 10. The two required conditions are satisfied.

Calculations (C):

$$(\hat{p}_1 - \hat{p}_2) \pm (z \text{ critical value})\sqrt{\frac{\hat{p}_1(1-\hat{p}_1)}{n_1} + \frac{\hat{p}_2(1-\hat{p}_2)}{n_2}}$$

$$(0.057 - 0.097) \pm 2.58\sqrt{\frac{0.057(1-0.057)}{400} + \frac{0.097(1-0.097)}{400}}$$

$$-0.040 \pm 2.58(0.018798)$$

$$-0.040 \pm 0.048498$$

$$(-0.0885, 0.0085)$$

Communicate Results (C):

Interpret confidence interval: We are 99% confident that the actual difference in the proportion of high school graduates who were unemployed in 2008 and high school graduates who were unemployed in 2009 is somewhere between –0.0885 and 0.0085. Because the endpoints of the confidence interval have opposite signs, zero is included in the interval, and there may be no difference in the proportion of high school graduates that were unemployed in 2008 and the proportion that were unemployed in 2009.

Interpret confidence level: The method used to construct this interval estimate is successful in capturing the actual value of the difference in population proportions 99% of the time.

(b) Wider, because the confidence level in part (a) is greater than the confidence level in the previous exercise, and the sample sizes are smaller.

Section 11.2 Exercise Set 1

11.11: (a) $H_0: p_1 - p_2 = 0$ versus $H_a: p_1 - p_2 > 0$, where p_1 is the actual proportion of workers who work over 40 hours per week who get less than 7 hours of sleep, and p_2 is the actual proportion of workers who work between 35-40 hours per week who get less than 7 hours of sleep.

(b) Yes, because there are more than 10 successes (those who get less than 7 hours of sleep) and 10 failures (those who do not get less than 7 hours of sleep) in each sample.

(c) The test statistic is $z = 3.63$ and the P-value = 0.000. Because the P-value of 0.000 is less than α of 0.01, we reject H_0.

(d) There is convincing evidence that the proportion of workers who usually get less than 7 hours of sleep a night is higher for those who work more than 40 hours per week than for those who work between 35 and 40 hours per week.

11.12: Using the 5-step process (HMC^3):

Hypotheses (H): We want to determine if there is evidence that the proportion of guests who reserve a room online who are satisfied is higher than the proportion of guests who reserve a room using a telephone system who are satisfied. Therefore, the population characteristics of interest are p_T, the proportion of guests who reserve rooms using a telephone system who are satisfied, and p_O, the proportion of guests who reserve rooms online who are satisfied. Test the hypotheses $H_0: p_T - p_O = 0$ versus $H_a: p_T - p_O < 0$.

Method (M): Because the answers to the four key questions are (Q) hypothesis testing, (S) sample data, (T) one categorical variable, and (N) two samples, we can consider a large-sample test for a difference in population proportions. The test statistic is: $z=\frac{\hat{p}_T-\hat{p}_O}{\sqrt{\frac{\hat{p}_C(1-\hat{p}_C)}{n_T}+\frac{\hat{p}_C(1-\hat{p}_C)}{n_O}}}$, and we will use a significance level of $\alpha=0.05$.

Check (C): We are told that the samples are independent and randomly selected. In addition, the sample of those who reserved a room using a telephone system contains 57 successes and 23 failures, and the sample of those who reserved a room online contains 50 successes and 10 failures; these four values are all at least 10. The two required conditions are satisfied.

Calculations (C): The sample proportions are $\hat{p}_T=\frac{57}{80}=0.7125$ and $\hat{p}_O=\frac{50}{60}=0.8333$. The combined proportion is $\hat{p}_C=\frac{n_T\hat{p}_T+n_O\hat{p}_O}{n_T+n_O}=\frac{(80)(0.7125)+(60)(0.8333)}{80+60}=0.7643$. Therefore, the test statistic is

$$z=\frac{\hat{p}_T-\hat{p}_O}{\sqrt{\frac{\hat{p}_C(1-\hat{p}_C)}{n_T}+\frac{\hat{p}_C(1-\hat{p}_C)}{n_O}}}=\frac{0.7125-0.8333}{\sqrt{\frac{0.7643(1-0.7643)}{80}+\frac{0.7643(1-0.7643)}{60}}}=-1.667$$

. This is a lower-tailed test (the inequality in H_a is <), so the P-value is the area under the z curve and to the left of -1.667. Therefore, the P-value is $P(z\le -1.667)=0.048$.

Communicate Results (C): Because the P-value of 0.048 is less than the selected significance level of $\alpha=0.05$, we reject the null hypothesis. There is convincing evidence that the proportion of guests who are satisfied is higher for those who reserve a room online.

11.13: Using the 5-step process (HMC^3):

Hypotheses (H): We want to determine if there is convincing evidence that the proportion of college graduates who experience sunburn is higher than the proportion of those without a high school degree who experience sunburn. Therefore, the population characteristics of interest are p_1, the proportion of college graduates who experience sunburn, and p_2, the proportion of those without a high school degree who experience sunburn. Test the hypotheses $H_0: p_1 - p_2 = 0$ versus $H_a: p_1 - p_2 > 0$.

Method (M): Because the answers to the four key questions are (Q) hypothesis testing, (S) sample data, (T) one categorical variable, and (N) two samples, we can consider a large-sample test for a difference in population proportions. The test statistic is: $z = \dfrac{\hat{p}_1 - \hat{p}_2}{\sqrt{\dfrac{\hat{p}_C(1-\hat{p}_C)}{n_1} + \dfrac{\hat{p}_C(1-\hat{p}_C)}{n_2}}}$, and we will use a significance level of $\alpha = 0.01$.

Check (C): We are told that the samples are independent and randomly selected. In addition, $n_1\hat{p}_1 = (200)(0.43) = 86$, $n_1(1-\hat{p}_1) = (200)(1-0.43) = 114$, $n_2\hat{p}_2 = (200)(0.25) = 50$, and $n_2(1-\hat{p}_2) = (200)(1-0.25) = 150$, which are all at least 10. The two required conditions are satisfied.

Calculations (C): The sample proportions are $\hat{p}_1 = 0.43$ and $\hat{p}_2 = 0.25$. The combined proportion is $\hat{p}_C = \dfrac{n_1\hat{p}_1 + n_2\hat{p}_2}{n_1 + n_2} = \dfrac{(200)(0.43) + (200)(0.25)}{200 + 200} = 0.34$.

Therefore, the test statistic is

$$z = \frac{\hat{p}_1 - \hat{p}_2}{\sqrt{\dfrac{\hat{p}_C(1-\hat{p}_C)}{n_1} + \dfrac{\hat{p}_C(1-\hat{p}_C)}{n_2}}} = \frac{0.43 - 0.25}{\sqrt{\dfrac{0.34(1-0.34)}{200} + \dfrac{0.34(1-0.34)}{200}}} = 3.800.$$

This is an upper-tailed test (the inequality in H_a is >), so the P-value is the area under the z curve and to the right of 3.800. Therefore, the P-value is $P(z \geq 3.800) \approx 0$.

Communicate Results (C): Because the P-value of approximately 0 is less than the selected significance level of $\alpha = 0.01$, we reject the null hypothesis. There is convincing evidence that the proportion experiencing sunburn is higher for college graduates than for those without a high school degree.

Section 11.2 Additional Exercises

11.17: Using the 5-step process (HMC³):

Hypotheses (H): We want to determine if there is convincing evidence that the proportion of young adults who think their parents would provide financial support and the proportion of parents who say they would provide financial support are different. Therefore, the population characteristics of interest are p_1, the proportion of all young adults who think their parents would provide financial support, and p_2, the proportion of parents of young adults who would provide financial support. Test the hypotheses $H_0: p_1 - p_2 = 0$ versus $H_a: p_1 - p_2 \neq 0$.

Method (M): Because the answers to the four key questions are (Q) hypothesis testing, (S) sample data, (T) one categorical variable, and (N) two samples, we can consider a large-sample test for a difference in population proportions. The test statistic is: $z = \dfrac{\hat{p}_1 - \hat{p}_2}{\sqrt{\dfrac{\hat{p}_C(1-\hat{p}_C)}{n_1} + \dfrac{\hat{p}_C(1-\hat{p}_C)}{n_2}}}$, and we will use a significance level of $\alpha = 0.05$.

Check (C): We are told that the samples are independent and randomly selected. In addition, $n_1\hat{p}_1 = (600)(0.41) = 246$, $n_1(1-\hat{p}_1) = (600)(1-0.41) = 354$, $n_2\hat{p}_2 = (300)(0.43) = 129$, and $n_2(1-\hat{p}_2) = (300)(1-0.43) = 171$, which are all at least 10. The two required conditions are satisfied.

Calculations (C): The sample proportions are $\hat{p}_1 = 0.41$ and $\hat{p}_2 = 0.43$. The combined proportion is $\hat{p}_C = \dfrac{n_1\hat{p}_1 + n_2\hat{p}_2}{n_1 + n_2} = \dfrac{(600)(0.41) + (300)(0.43)}{600 + 300} = 0.417$.

Therefore, the test statistic is

$$z = \frac{\hat{p}_1 - \hat{p}_2}{\sqrt{\dfrac{\hat{p}_C(1-\hat{p}_C)}{n_1} + \dfrac{\hat{p}_C(1-\hat{p}_C)}{n_2}}} = \frac{0.41 - 0.43}{\sqrt{\dfrac{0.417(1-0.417)}{600} + \dfrac{0.417(1-0.417)}{300}}} = -0.574.$$

This is a two-tailed test (the inequality in H_a is $\neq$), so the P-value is the total area under the z curve and to the left of -0.574 and to the right of 0.574. Therefore, by the symmetry of the standard normal curve, the P-value is $2 \cdot P(z \leq -0.574) = 0.566$.

Communicate Results (C): Because the P-value of 0.566 is greater than the selected significance level of $\alpha = 0.05$, we fail to reject the null hypothesis. There is not convincing evidence of a difference between the proportion of young adults who think that their parents would provide financial support for marriage and the proportion of parents who say they would provide financial support for marriage.

11.19: Since the data given are population characteristics, an inference procedure is not applicable. It is *known* that the rate of Lou Gehrig's disease among soldiers sent to the war is higher than for those not sent to the war.

Are You Ready To Move On? Chapter 11 Review Exercises

11.21: Using the 5-step process (HMC^3):

Hypotheses (H): We want to determine if there is convincing evidence that the proportion of teens that approve of banning cell phone use and texting while driving is less than the proportion of parents of teens who approve. Therefore, the population characteristics of interest are p_1, the proportion of all teens who approve of banning cell phone use and texting while driving, and p_2, the proportion of all parents of teens who approve of banning cell phone use and texting while driving. Test the hypotheses $H_0: p_1 - p_2 = 0$ versus $H_a: p_1 - p_2 < 0$.

Method (M): Because the answers to the four key questions are (Q) hypothesis testing, (S) sample data, (T) one categorical variable, and (N) two samples, we can consider a large-sample test for a difference in population proportions. The test statistic is: $z = \dfrac{\hat{p}_1 - \hat{p}_2}{\sqrt{\dfrac{\hat{p}_C(1-\hat{p}_C)}{n_1} + \dfrac{\hat{p}_C(1-\hat{p}_C)}{n_2}}}$, and we will use a significance level of $\alpha = 0.05$.

Check (C): We are told to assume that the samples are representative of the two populations. In addition, $n_1\hat{p}_1 = (600)(0.74) = 444$, $n_1(1-\hat{p}_1) = (600)(1-0.74) = 156$, $n_2\hat{p}_2 = (400)(0.95) = 380$, and $n_2(1-\hat{p}_2) = (400)(1-0.95) = 20$, which are all at least 10. The two required conditions are satisfied.

Calculations (C): The sample proportions are $\hat{p}_1 = 0.74$ and $\hat{p}_2 = 0.95$. The combined proportion is $\hat{p}_C = \dfrac{n_1\hat{p}_1 + n_2\hat{p}_2}{n_1 + n_2} = \dfrac{(600)(0.74) + (400)(0.95)}{600 + 400} = 0.824$.

Therefore, the test statistic is

$$z=\frac{\hat{p}_1-\hat{p}_2}{\sqrt{\frac{\hat{p}_C(1-\hat{p}_C)}{n_1}+\frac{\hat{p}_C(1-\hat{p}_C)}{n_2}}}=\frac{0.74-0.95}{\sqrt{\frac{0.824(1-0.824)}{600}+\frac{0.824(1-0.824)}{400}}}=-8.543.$$

This is a lower-tailed test (the inequality in H_a is $<$), so the P-value is the area under the z curve and to the left of –8.543. Therefore, the P-value is $P(z \le -8.543) \approx 0$.

Communicate Results (C): Because the P-value of approximately 0 is less than the selected significance level of $\alpha = 0.05$, we reject the null hypothesis. There is convincing evidence that the proportion of teens who approve of the proposed laws is less than the proportion of parents of teens who approve.

11.23: (a) Using the 5-step process (HMC^3):

Hypotheses (H): We want to determine if there is convincing evidence that the proportion of those who think that newspapers are boring is different for teenage girls and boys. Therefore, the population characteristics of interest are p_G, the proportion of all teenage girls who think that newspapers are boring, and p_B, the proportion of all teenage boys who think that newspapers are boring. Test the hypotheses $H_0: p_G - p_B = 0$ versus $H_a: p_G - p_B \neq 0$.

Method (M): Because the answers to the four key questions are (Q) hypothesis testing, (S) sample data, (T) one categorical variable, and (N) two samples, we can consider a large-sample test for a difference in population proportions. The test statistic is: $z=\frac{\hat{p}_G-\hat{p}_B}{\sqrt{\frac{\hat{p}_C(1-\hat{p}_C)}{n_G}+\frac{\hat{p}_C(1-\hat{p}_C)}{n_B}}}$, and we will use a significance level of $\alpha = 0.05$.

Check (C): We are told that the samples are representative of the two populations. In addition, $n_G\hat{p}_G = (58)(0.41) = 23.78$, $n_G(1-\hat{p}_G) = (58)(1-0.41) = 34.22$, $n_B\hat{p}_B = (41)(0.44) = 18.04$, and $n_B(1-\hat{p}_B) = (41)(1-0.44) = 22.96$, which are all at least 10. The two required conditions are satisfied.

Calculations (C): The sample proportions are $\hat{p}_G = 0.41$ and $\hat{p}_B = 0.44$. The combined proportion is $\hat{p}_C = \frac{n_G\hat{p}_G + n_B\hat{p}_B}{n_G + n_B} = \frac{(58)(0.41)+(41)(0.44)}{58+41} = 0.422$.

Therefore, the test statistic is

$$z = \frac{\hat{p}_G - \hat{p}_B}{\sqrt{\frac{\hat{p}_C(1-\hat{p}_C)}{n_G} + \frac{\hat{p}_C(1-\hat{p}_C)}{n_B}}} = \frac{0.41-0.44}{\sqrt{\frac{0.422(1-0.422)}{58} + \frac{0.422(1-0.422)}{41}}} = -0.298.$$

This is a two-tailed test (the inequality in H_a is $\neq$), so the *P*-value is the total area under the *z* curve and to the left of –0.298 and to the right of 0.298. Therefore, using the symmetry of the normal curve, the *P*-value is $2 \cdot P(z \leq -0.298) = 0.766$.

Communicate Results (C): Because the *P*-value of 0.766 is greater than the selected significance level of $\alpha = 0.05$, we fail to reject the null hypothesis. There is not convincing evidence that the proportion of those who think that newspapers are boring is different for teenage girls and boys.

(b) Using the 5-step process (HMC^3):

Hypotheses (H): We want to determine if there is convincing evidence that the proportion of those who think that newspapers are boring is different for teenage girls and boys. Therefore, the population characteristics of interest are p_G, the proportion of all teenage girls who think that newspapers are boring, and p_B, the proportion of all teenage boys who think that newspapers are boring. Test the hypotheses $H_0: p_G - p_B = 0$ versus $H_a: p_G - p_B \neq 0$.

Method (M): Because the answers to the four key questions are (Q) hypothesis testing, (S) sample data, (T) one categorical variable, and (N) two samples, we can consider a large-sample test for a difference in population proportions. The test statistic is: $z = \frac{\hat{p}_G - \hat{p}_B}{\sqrt{\frac{\hat{p}_C(1-\hat{p}_C)}{n_G} + \frac{\hat{p}_C(1-\hat{p}_C)}{n_B}}}$, and we will use a significance level of $\alpha = 0.05$.

Check (C): We are told that the samples are representative of the two populations. In addition, $n_G\hat{p}_G = (2,000)(0.41) = 820$, $n_G(1-\hat{p}_G) = (2,000)(1-0.41) = 1,180$, $n_B\hat{p}_B = (2,500)(0.44) = 1,100$, and

$n_B(1-\hat{p}_B) = (2{,}500)(1-0.44) = 1{,}400$, which are all at least 10. The two required conditions are satisfied.

Calculations (C): The sample proportions are $\hat{p}_G = 0.41$ and $\hat{p}_B = 0.44$. The combined proportion is

$$\hat{p}_C = \frac{n_G\hat{p}_G + n_B\hat{p}_B}{n_G + n_B} = \frac{(2{,}000)(0.41)+(2{,}500)(0.44)}{2{,}000+2{,}500} = 0.427.$$ Therefore, the test statistic is

$$z = \frac{\hat{p}_G - \hat{p}_B}{\sqrt{\frac{\hat{p}_C(1-\hat{p}_C)}{n_G} + \frac{\hat{p}_C(1-\hat{p}_C)}{n_B}}} = \frac{0.41-0.44}{\sqrt{\frac{0.427(1-0.427)}{2{,}000} + \frac{0.427(1-0.427)}{2{,}500}}} = -2.022.$$

This is a two-tailed test (the inequality in H_a is $\neq$), so the P-value is the total area under the z curve and to the left of -2.022 and to the right of 2.022. Therefore, using the symmetry of the normal curve, the P-value is $2 \cdot P(z \leq -2.022) = 0.043$.

Communicate Results (C): Because the P-value of 0.043 is less than the selected significance level of $\alpha = 0.05$, we reject the null hypothesis. There is convincing evidence that the proportion of those who think that newspapers are boring is different for teenage girls and boys.

(c) Assuming that the population proportions are equal, you are much less likely to get a difference in sample proportions as large as the one given when the samples are very large than when the samples are relatively small.

Chapter 12 Asking and Answering Questions About A Population Mean

Section 12.1 Exercise Set 1

12.1: (a) $\mu_{\bar{x}} = \mu = 100,\ \sigma_{\bar{x}} = \frac{\sigma}{\sqrt{n}} = \frac{10}{\sqrt{9}} = 3.333$

(b) $\mu_{\bar{x}} = \mu = 100,\ \sigma_{\bar{x}} = \frac{\sigma}{\sqrt{n}} = \frac{10}{\sqrt{15}} = 2.582$

(c) $\mu_{\bar{x}} = \mu = 100,\ \sigma_{\bar{x}} = \frac{\sigma}{\sqrt{n}} = \frac{10}{\sqrt{36}} = 1.667$

(d) $\mu_{\bar{x}} = \mu = 100,\ \sigma_{\bar{x}} = \frac{\sigma}{\sqrt{n}} = \frac{10}{\sqrt{50}} = 1.414$

(e) $\mu_{\bar{x}} = \mu = 100,\ \sigma_{\bar{x}} = \frac{\sigma}{\sqrt{n}} = \frac{10}{\sqrt{100}} = 1$

(f) $\mu_{\bar{x}} = \mu = 100,\ \sigma_{\bar{x}} = \frac{\sigma}{\sqrt{n}} = \frac{10}{\sqrt{400}} = 0.5$

12.2: The sampling distribution of $\bar{x}$ will be approximately normal for the sample sizes in parts (c)–(f), since those sample sizes are all greater than or equal to 30.

12.3:

What You Know	How You Know It
The sampling distribution of $\bar{x}$ is centered at the actual (but unknown) value of the population mean	Rule 1 states that the mean of the sampling distribution of $\bar{x}$ is equal to the population mean, $\mu_{\bar{x}} = \mu$.
An estimate of the standard deviation of $\bar{x}$, which describes how much the $\bar{x}$ values spread out around the population mean μ,is 0.455.	Rule 2 states that the standard deviation of the sampling distribution of $\bar{x}$ is $\sigma_{\bar{x}} = \frac{\sigma}{\sqrt{n}}$, which is approximately equal to $\frac{s}{\sqrt{n}} = \frac{6.62}{\sqrt{212}} = 0.455$.
The sampling distribution of $\bar{x}$ is approximately normal.	Rule 4 states that the sampling distribution of $\bar{x}$ is approximately normal if the sample size, n, is large. In this case, $n = 212$, which is larger than 30.

12.4: The quantity μ is the population mean, while $\mu_{\bar{x}}$ is the mean of the sampling distribution of $\bar{x}$. It is the mean value of $\bar{x}$ for all possible random samples of size *n*.

12.5: (a) $\mu_{\bar{x}} = \mu = 0.5$, $\sigma_{\bar{x}} = \dfrac{\sigma}{\sqrt{n}} = \dfrac{0.289}{\sqrt{16}} = 0.07225$

(b) $\mu_{\bar{x}} = \mu = 0.5$, $\sigma_{\bar{x}} = \dfrac{\sigma}{\sqrt{n}} = \dfrac{0.289}{\sqrt{50}} = 0.040871$. The sampling distribution of $\bar{x}$ is approximately normal because $n \geq 30$.

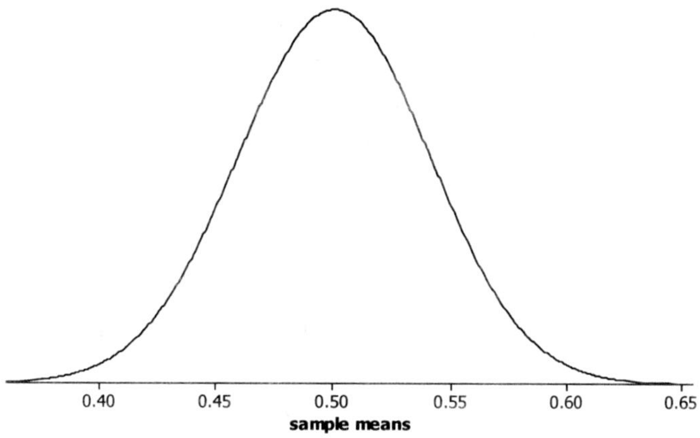

Section 12.1 Additional Exercises

12.11: (a) $\mu_{\bar{x}} = \mu = 200$, $\sigma_{\bar{x}} = \dfrac{\sigma}{\sqrt{n}} = \dfrac{15}{\sqrt{12}} = 4.330$

(b) $\mu_{\bar{x}} = \mu = 200$, $\sigma_{\bar{x}} = \dfrac{\sigma}{\sqrt{n}} = \dfrac{15}{\sqrt{20}} = 3.354$

(c) $\mu_{\bar{x}} = \mu = 200$, $\sigma_{\bar{x}} = \dfrac{\sigma}{\sqrt{n}} = \dfrac{15}{\sqrt{25}} = 3$

(d) $\mu_{\bar{x}} = \mu = 200$, $\sigma_{\bar{x}} = \dfrac{\sigma}{\sqrt{n}} = \dfrac{15}{\sqrt{40}} = 2.372$

(e) $\mu_{\bar{x}} = \mu = 200$, $\sigma_{\bar{x}} = \dfrac{\sigma}{\sqrt{n}} = \dfrac{15}{\sqrt{90}} = 1.581$

(f) $\mu_{\bar{x}} = \mu = 200$, $\sigma_{\bar{x}} = \dfrac{\sigma}{\sqrt{n}} = \dfrac{15}{\sqrt{300}} = 0.866$

12.13: The quantity $\bar{x}$ is the mean of a single sample, while $\mu_{\bar{x}}$ is the mean of the sampling distribution of $\bar{x}$. It is the mean value of $\bar{x}$ for all possible random samples of size *n*.

12.15: (a) We know that $\mu_{\bar{x}} = \mu = 65$ mm and $\sigma_{\bar{x}} = \dfrac{\sigma}{\sqrt{n}} = \dfrac{5}{\sqrt{25}} = 1$ mm. Also, since we're told that the distribution of all interpupillary distances is normal we know that the sampling distribution of $\bar{x}$ is normal. So the probability that the sample mean distance is between 64 and 67 mm is $P(64 \le \bar{x} \le 67) = P\left(\dfrac{64-65}{1} \le z \le \dfrac{67-65}{1}\right) = P(-1 \le z \le 2) = 0.8186$. The probability that the sample mean distance is at least 68 mm is $P(\bar{x} \ge 68) = P\left(z \ge \dfrac{68-65}{1}\right) = P(z \ge 3) = 0.00135$.

(b) We know that $\mu_{\bar{x}} = \mu = 65$ mm and $\sigma_{\bar{x}} = \dfrac{\sigma}{\sqrt{n}} = \dfrac{5}{\sqrt{100}} = 0.5$ mm. Also, since *n* = 100, which is greater than 30, the sampling distribution of $\bar{x}$ is approximately normal. So the approximate probability that the sample mean distance is between 64 and 67 mm is $P(64 \le \bar{x} \le 67) = P\left(\dfrac{64-65}{0.5} \le z \le \dfrac{67-65}{0.5}\right) = P(-2 \le z \le 4) = 0.9772$. The approximate probability that the sample mean distance is at least 68 mm is $P(\bar{x} \ge 68) = P\left(z \ge \dfrac{68-65}{0.5}\right) = P(z \ge 6) \approx 0$.

12.17: We know that $\mu_{\bar{x}} = \mu = 0.5$ inches and $\sigma_{\bar{x}} = \dfrac{\sigma}{\sqrt{n}} = \dfrac{0.02}{\sqrt{36}} = 0.003333$, so the probability of observing an $\bar{x}$ between 0.49 inches and 0.51 inches is $P(0.49 < \bar{x} < 0.51) = P\left(\dfrac{0.49-0.5}{0.003333} \le z \le \dfrac{0.51-0.5}{0.003333}\right) = P(-3.0003 \le z \le 3.0003) = 0.9973$. Therefore, the probability that the manufacturing line will be shut down unnecessarily is $1 - 0.9973 = 0.0027$.

Section 12.2 Exercise Set 1

12.19: (a) Using technology or the t table, 90%.

(b) Using technology or the t table, 95%.

(c) The percent outside the interval from –2.80 to 2.80 with 24 degrees of freedom is the complement of the percent inside the interval. Using technology or the t table, the percent inside the interval is 99%. Therefore, the percent outside the interval is $1-0.99=0.01$, or 1%.

(d) Using technology or the t table, 5%.

12.20: (a) Using technology or the t table, the t critical value is 2.12.

(b) Using technology or the t table, the t critical value is 2.81.

(c) Using technology or the t table, the t critical value is 1.78.

12.21: (a) Since the sample mean, $\bar{x}$, is centered in the confidence interval, its value is the average of the endpoints of the interval. Therefore, $\bar{x}=\dfrac{114.4+115.6}{2}=115.0$.

Note that using the wider interval also results in the same sample mean: $\bar{x}=\dfrac{114.1+115.9}{2}=115.0$.

(b) The 99% confidence interval is wider than the 90% confidence interval, because the t critical value for 99% confidence is greater than the t critical value for 90% confidence. Therefore, the 90% interval is (114.4, 115.6) and the 99% interval is (114.1, 115.9).

12.22: (a) The fact that the mean is much greater than the median suggests that the distribution of times spent volunteering in the sample was positively skewed.

(b) With the sample mean much greater than the sample median, and with the sample regarded as representative of the population, it seems very likely that the population is strongly positively skewed and, therefore, not normally distributed.

(c) Since $n=1,086\geq 30$, the sample size is large enough for the one-sample t confidence interval to be appropriate, even though the population distribution is not approximately normal.

(d) Using the five-step process (EMC^3):

Estimate (E): The mean time spent volunteering for the population of parents of school-aged children, μ, will be estimated.

Method (M): Because the answers to the four key questions are (Q) estimation, (S) sample data, (T) one numerical variable, and (N) one sample, a one-sample *t* confidence interval for a population mean will be considered. A confidence level of 98% was specified.

Check (C): There are two conditions that need to be met in order for the one-sample *t* confidence interval to be appropriate. The sample was selected to be representative of the population of parents of school-aged children. Additionally, the sample size, $n = 1,086$, is significantly larger than 30, so it is appropriate to use the *t* confidence interval. Both conditions have been satisfied.

Calculate (C): From Appendix Table 3, the *t* critical value is 2.33 (from the *z* critical value row because $df = 1,086 - 1 = 1,085 > 120$, the largest number of degrees of freedom from the table). The 98% confidence interval for μ is then:

$$\bar{x} \pm (t \text{ critical value}) \frac{s}{\sqrt{n}}$$

$$5.6 \pm 2.33 \frac{5.2}{\sqrt{1086}}$$

$$5.6 \pm 2.33(0.15779)$$

$$5.6 \pm 0.3677$$

$$(5.232, 5.968)$$

Communicate Results (C):

Interpret Confidence Interval: Based on this sample, you can be 98% confident that the mean time spent volunteering for the population of parents of school-age children is between 5.232 and 5.968 hours.

Interpret Confidence Level: The method used to construct this interval estimate is successful in capturing the actual value of the population mean about 98% of the time.

12.23: Using the five-step process (EMC^3):

Estimate (E): The mean time spent studying for the introductory psychology exam, μ, will be estimated.

Method (M): Because the answers to the four key questions are (Q) estimation, (S) sample data, (T) one numerical variable, and (N) one sample, a one-sample *t* confidence interval for a population mean will be considered. A confidence level of 95% was specified.

Check (C): There are two conditions that need to be met in order for the one-sample *t* confidence interval to be appropriate. The sample was selected to be representative of the population of students taking introductory psychology at the university. Additionally, the sample size, $n = 411$, is significantly larger than 30, so it is appropriate to use the *t* confidence interval. Both conditions have been satisfied.

Calculate (C): From Appendix Table 3, the *t* critical value is 1.96 (from the *z* critical value row because $df = 411 - 1 = 410 > 120$, the largest number of degrees of freedom from the table). The 95% confidence interval for μ is then:

$$\bar{x} \pm (t \text{ critical value})\frac{s}{\sqrt{n}}$$

$$7.74 \pm 1.96\frac{3.40}{\sqrt{411}}$$

$$7.74 \pm 1.96(0.16771)$$

$$7.74 \pm 0.3287$$

$$(7.411, 8.069)$$

Communicate Results (C):

Interpret Confidence Interval: Based on this sample, you can be 95% confident that the mean time spent studying for the final exam is between 7.411 and 8.069 hours.

Interpret Confidence Level: The method used to construct this interval estimate is successful in capturing the actual value of the population mean about 95% of the time.

12.24: Using the five-step process (EMC^3):

Estimate (E): The mean percentage of study time that occurs in the 24 hours prior to the final exam, μ, will be estimated.

Method (M): Because the answers to the four key questions are (Q) estimation, (S) sample data, (T) one numerical variable, and (N) one sample, a one-sample t confidence interval for a population mean will be considered. A confidence level of 90% was specified.

Check (C): There are two conditions that need to be met in order for the one-sample t confidence interval to be appropriate. The sample was selected to be representative of the population of students taking introductory psychology at the university. Additionally, the sample size, $n = 411$, is significantly larger than 30, so it is appropriate to use the t confidence interval. Both conditions have been satisfied.

Calculate (C): From Appendix Table 3, the t critical value is 1.645 (from the z critical value row because $df = 411 - 1 = 410 > 120$, the largest number of degrees of freedom from the table). The 90% confidence interval for μ is then:

$$\bar{x} \pm (t \text{ critical value})\frac{s}{\sqrt{n}}$$

$$43.18 \pm 1.645\frac{21.46}{\sqrt{411}}$$

$$43.18 \pm 1.645(1.05854)$$

$$43.18 \pm 1.7413$$

$$(41.439, 44.921)$$

Communicate Results (C):

Interpret Confidence Interval: Based on this sample, you can be 90% confident that the mean percentage of study time that occurs in the 24 hours prior to the final exam is between 41.439% and 44.921%.

Interpret Confidence Level: The method used to construct this interval estimate is successful in capturing the actual value of the population mean about 95% of the time.

12.25: Estimate the population standard deviation using

$$\frac{\text{sample range}}{4} = \frac{700-50}{4} = 162.5.$$ Therefore, the sample size needed is

$$n = \left(\frac{1.96\sigma}{M}\right)^2 = \left(\frac{1.96(162.5)}{10}\right)^2 = 1014.42,$$ so a sample size of 1015 is needed.

12.26: (a) This could happen if the sample distribution was strongly positively skewed.

(b) No. Since the sample distribution is strongly skewed to the right, it is quite unlikely that the variable anticipated Halloween expense is approximately normally distributed.

(c) Yes, since the sample size of $n = 1{,}000$ is much larger than 30.

(d) Using the five-step process (EMC^3):

Estimate (E): The mean anticipated Halloween expense for Canadian residents, μ, will be estimated.

Method (M): Because the answers to the four key questions are (Q) estimation, (S) sample data, (T) one numerical variable, and (N) one sample, a one-sample t confidence interval for a population mean will be considered. A confidence level of 99% was specified.

Check (C): There are two conditions that need to be met in order for the one-sample t confidence interval to be appropriate. The sample was randomly selected from the population of Canadian residents. Additionally, the sample size, $n = 1{,}000$, is significantly larger than 30, so it is appropriate to use the t confidence interval. Both conditions have been satisfied.

Calculate (C): From Appendix Table 3, the t critical value is 2.58 (from the z critical value row because $df = 1{,}000 - 1 = 999 > 120$, the largest number of degrees of freedom from the table). The 99% confidence interval for μ is then:

$$\bar{x} \pm (t \text{ critical value})\frac{s}{\sqrt{n}}$$

$$46.65 \pm 2.58\frac{83.70}{\sqrt{1000}}$$

$$46.65 \pm 2.58(2.64683)$$

$$46.65 \pm 6.82881$$

$$(39.821, 53.479)$$

Communicate Results (C):

Interpret Confidence Interval: Based on this sample, you can be 99% confident that the mean anticipated Halloween expense for Canadian residents is between \$39.821 and \$53.479.

Interpret Confidence Level: The method used to construct this interval estimate is successful in capturing the actual value of the population mean about 99% of the time.

Section 12.2 Additional Exercises

12.35: (a) Using technology, the t critical value is 2.14. Using the t table, the t critical value is 2.15.

(b) Using technology or the t table, the t critical value is 2.86.

(c) Using technology or the t table, the t critical value is 1.71.

12.37: Using the five-step process (EMC^3):

Estimate (E): The mean number of months elapsed since the last visit to a dentist for the population of students participating in the program, μ, will be estimated.

Method (M): Because the answers to the four key questions are (Q) estimation, (S) sample data, (T) one numerical variable, and (N) one sample, a one-sample t confidence interval for a population mean will be considered. A confidence level of 95% was specified.

Check (C): There are two conditions that need to be met in order for the one-sample t confidence interval to be appropriate. The sample is representative of all students participating in the free checkup program. Additionally, the sample size, $n = 5$, is small, we must consider whether it is reasonable to think that the distribution of the number of months since last dental visit is approximately normal. A normal probability plot of the sample data is shown below.

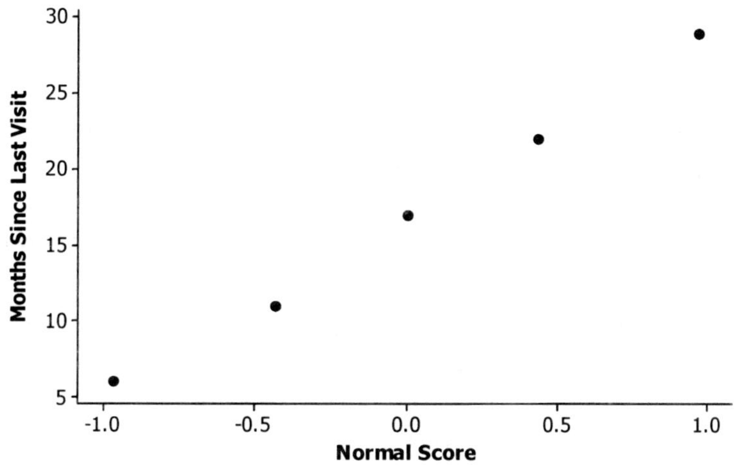

Based on the linearity of the plot, it seems plausible that the population distribution is approximately normal. Both necessary conditions appear to be satisfied.

Calculate (C): From Appendix Table 3, the t critical value for 95% confidence with 4 degrees of freedom is 2.78. The 95% confidence interval for μ is then:

$$\bar{x} \pm (t \text{ critical value})\frac{s}{\sqrt{n}}$$

$$17 \pm 2.78\frac{9.03}{\sqrt{5}}$$

$$17 \pm 2.78(4.03834)$$

$$17 \pm 11.2266$$

$$(5.7734, 28.2266)$$

Communicate Results (C):

Interpret Confidence Interval: Based on this sample, you can be 95% confident that the mean number of months elapsed since the last visit to a dentist for the population of students participating in the program is between 5.773 and 28.227 months.

Interpret Confidence Level: The method used to construct this interval estimate is successful in capturing the actual value of the population mean about 95% of the time.

12.39: Using the five-step process (EMC^3):

Estimate (E): The mean number of partners on a mating flight for queen honeybees, μ, will be estimated.

Method (M): Because the answers to the four key questions are (Q) estimation, (S) sample data, (T) one numerical variable, and (N) one sample, a one-sample t confidence interval for a population mean will be considered. A confidence level of 95% was specified.

Check (C): There are two conditions that need to be met in order for the one-sample t confidence interval to be appropriate. The sample is representative of the population of queen honeybees. Additionally, the sample size, $n = 30$, is large enough for us to use the t confidence interval. Both necessary conditions are satisfied.

Calculate (C): From Appendix Table 3, the t critical value for 95% confidence with 29 degrees of freedom is 2.05. The 95% confidence interval for μ is then:

$$\bar{x} \pm (t \text{ critical value})\frac{s}{\sqrt{n}}$$

$$4.6 \pm 2.05\frac{3.47}{\sqrt{30}}$$

$$4.6 \pm 2.05(0.633532)$$

$$4.6 \pm 1.29874$$

$$(3.301, 5.899)$$

Communicate Results (C):

Interpret Confidence Interval: Based on this sample, you can be 95% confident that the mean number of partners on a mating flight for queen honeybees is between 3.301 and 5.899.

Interpret Confidence Level: The method used to construct this interval estimate is successful in capturing the actual value of the population mean about 95% of the time.

Section 12.3 Exercise Set 1

12.41: (a) With 8 degrees of freedom, $P\text{-value} = P(t \geq 2) = 0.040$.

(b) With 10 degrees of freedom, $P\text{-value} = P(t \leq -2.4) = 0.019$.

(c) With 21 degrees of freedom, $P\text{-value} = P(t \leq -4.2) = 0.000$.

(d) With 15 degrees of freedom, $P\text{-value} = 2 \cdot P(t \leq -1.6) = 0.130$.

12.42: Using the five-step process (HMC[3]):

Hypotheses (H): We want to know if the sample data provide convincing evidence that the mean heart rate after 15 minutes of Wii Bowling is different from the known mean heart rate after six minutes of walking on the treadmill. The population characteristic of interest is μ, the true mean heart rate after 15 minutes of Wii Bowling. Hypotheses: $H_0: \mu = 98$ minutes versus $H_a: \mu \neq 98$ minutes.

Method (M): Because the answers to the four key questions are (Q) hypothesis testing, (S) sample data, (T) one numerical variable, and (N) one sample, we can use a one-sample t test for a population mean. The test statistic for this test is $t = \dfrac{\bar{x} - 98}{\frac{s}{\sqrt{n}}}$, with $\text{df} = n - 1 = 14 - 1 = 13$. Use a significance level of $\alpha = 0.01$.

Check (C): We are told to assume that the sample of boys is representative of boys age 10 to 13. Additionally, we are also told that the distribution of heart rates after 15 minutes of Wii Bowling is approximately normal. Therefore, both of the necessary conditions have been satisfied.

Calculate (C): Compute the test statistic with $\bar{x} = 101$ bpm, $s = 15$ bpm, and $n = 14$. Therefore, $t = \dfrac{101 - 98}{\frac{15}{\sqrt{14}}} = 0.748$. This is a two-tailed test (the inequality in H_a is $\neq$), so the P-value is the area under the t curve with 13 degrees of freedom above t = 0.748 and below t = -0.748. Using the symmetry of the t distribution, $P\text{-value} = 2 \cdot P(t \geq 0.748) = 0.468$.

Communicate Results (C): Because the P-value of 0.468 is greater than the specified significance level of α = 0.01, we fail to reject H_0. There is not convincing evidence that the mean heart rate after 15 minutes of Wii Bowling is different from the known mean heart rate after six minutes of walking on the treadmill.

12.43: (a) Using the five-step process (HMC^3):

Hypotheses (H): We want to know if the sample data provide convincing evidence that the mean heart rate after 15 minutes of Wii Bowling is higher than the known mean resting heart rate for boys in this age group. The population characteristic of interest is μ, the true mean resting heart rate for boys in this age group. Hypotheses: $H_0 : \mu = 66$ bpm versus $H_a : \mu > 66$ bpm.

Method (M): Because the answers to the four key questions are (Q) hypothesis testing, (S) sample data, (T) one numerical variable, and (N) one sample, we can use a one-sample t test for a population mean. The test statistic for this test is $t = \dfrac{\bar{x} - 66}{\frac{s}{\sqrt{n}}}$, with $\text{df} = n - 1 = 14 - 1 = 13$. Use a significance level of $\alpha = 0.01$.

Check (C): We are told to assume that the sample of boys is representative of boys age 10 to 13. Additionally, we are also told that the distribution of heart rates after 15 minutes of Wii Bowling is approximately normal. Therefore, both of the necessary conditions have been satisfied.

Calculate (C): Compute the test statistic with $\bar{x} = 101$ bpm, $s = 15$ bpm, and $n = 14$. Therefore, $t = \dfrac{101 - 66}{\frac{15}{\sqrt{14}}} = 8.731$. This is an upper-tailed test (the inequality in H_a is >), so the P-value is the area under the t curve with 13 degrees of freedom above t = 8.731. Therefore, $P\text{-value} = P(t \geq 8.731) \approx 0$.

Communicate Results (C): Because the P-value of approximately 0 is smaller than the specified significance level of α = 0.01, we reject H_0. There is convincing evidence that the mean heart rate after 15 minutes of Wii Bowling is different from the known mean resting heart rate for boys of this age.

(b) Answers may vary. Our sample data do not provide convincing evidence that heart rates of boys ages 10 to 13 walking on a treadmill at 2.6 km/hr for 6 minutes is different from playing Wii Bowling for 15 minutes. However, we do have convincing evidence that the mean heart rate of boys in this age who play 15 minutes of Wii Bowling is higher than the known mean resting heart rate for boys in this age. Therefore, it seems as if it makes no difference whether boys in this age play Wii Bowling or walk on a treadmill in terms of the effect on heart rates.

12.44: Using the five-step process (HMC[3]):

Hypotheses (H): We want to know if the sample data provide convincing evidence that the mean May 2009 price of a Big Mac in Europe is greater than the reported U.S. price. The population characteristic of interest is μ, the true mean May 2009 price of a Big Mac in Europe. Hypotheses: $H_0 : \mu = 3.57$ versus $H_a : \mu > 3.57$.

Method (M): Because the answers to the four key questions are (Q) hypothesis testing, (S) sample data, (T) one numerical variable, and (N) one sample, we can use a one-sample t test for a population mean. The test statistic for this test is $t = \dfrac{\bar{x} - 3.57}{\frac{s}{\sqrt{n}}}$, with $\text{df} = n - 1 = 8 - 1 = 7$. Use a significance level of $\alpha = 0.05$.

Check (C): We are told it is reasonable to regard the sample as representative of European McDonald's restaurants. The sample of size $n = 8$ is small, so we must verify that the normality condition is reasonable. The boxplot of the sample data is shown below.

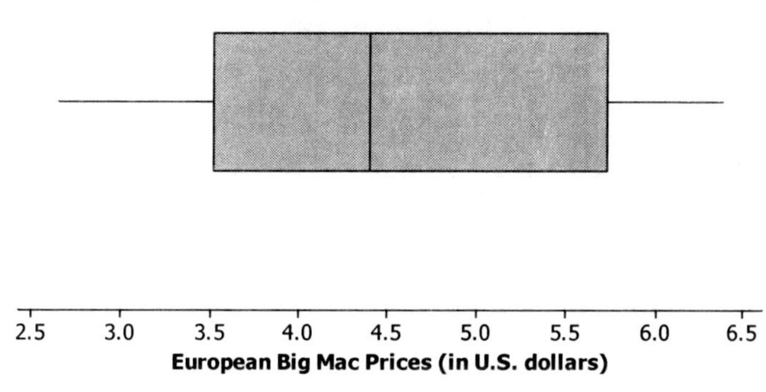

The boxplot is approximately symmetric and does not contain any outliers, so it is reasonable to think that the population distribution is at least approximately normal. Therefore, both of the necessary conditions have been satisfied.

Calculate (C): Compute the test statistic with $\bar{x} = 4.5425$, $s = 1.34672$, and $n = 8$. Therefore, $t = \frac{4.5425 - 3.57}{\frac{1.34672}{\sqrt{8}}} = 2.042$. This is an upper-tailed test (the inequality in H_a is >), so the P-value is the area under the t curve with 7 degrees of freedom above $t = 2.042$. Therefore, $P\text{-value} = P(t \geq 2.042) = 0.040$.

Communicate Results (C): Because the P-value of 0.040 is smaller than the specified significance level of $\alpha = 0.05$, we reject H_0. There is convincing evidence that the mean May 2009 price of a Big Mac in Europe is greater than the reported U.S. price.

12.45: (a) Using the five-step process (HMC^3):

Hypotheses (H): We want to know if the sample data provide convincing evidence that the mean time spent online by Canadians in a typical week is greater than 12.5 hours. The population characteristic of interest is μ, the true mean time spent online by Canadians in a typical week. Hypotheses: $H_0: \mu = 12.5$ hours versus $H_a: \mu > 12.5$ hours.

Method (M): Because the answers to the four key questions are (Q) hypothesis testing, (S) sample data, (T) one numerical variable, and (N) one sample, we can use a one-sample t test for a population mean. The test statistic for this test is $t = \frac{\bar{x} - 12.5}{\frac{s}{\sqrt{n}}}$, with $\text{df} = n - 1 = 1{,}000 - 1 = 999$. Use a significance level of $\alpha = 0.05$.

Check (C): We are told that the sample of Canadian Internet users was randomly selected. In addition, the sample of size n = 1,000 is much larger than 30. Therefore, both of the necessary conditions have been satisfied.

Calculate (C): Compute the test statistic with $\bar{x} = 12.7$ hours, $s = 5$ hours, and $n = 1,000$. Therefore, $t = \frac{12.7 - 12.5}{\frac{5}{\sqrt{1000}}} = 1.2649$. This is an upper-tailed test (the inequality in H_a is >), so the P-value is the area under the t curve with 999 degrees of freedom above $t = 1.2649$. Therefore, $P\text{-value} = P(t \geq 1.2649) = 0.103$.

Communicate Results (C): Because the P-value of 0.103 is larger than the specified significance level of $\alpha = 0.05$, we fail to reject H_0. There is not convincing evidence that the mean time spent online by Canadians in a typical week is greater than 12.5 hours.

(b) Using the five-step process (HMC^3):

Hypotheses (H): We want to know if the sample data provide convincing evidence that the mean time spent online by Canadians in a typical week is greater than 12.5 hours. The population characteristic of interest is μ, the true mean time spent online by Canadians in a typical week. Hypotheses: $H_0: \mu = 12.5$ hours versus $H_a: \mu > 12.5$ hours.

Method (M): Because the answers to the four key questions are (Q) hypothesis testing, (S) sample data, (T) one numerical variable, and (N) one sample, we can use a one-sample t test for a population mean. The test statistic for this test is $t = \frac{\bar{x} - 12.5}{\frac{s}{\sqrt{n}}}$, with $\text{df} = n - 1 = 1,000 - 1 = 999$. Use a significance level of $\alpha = 0.05$.

Check (C): We are told that the sample of Canadian Internet users was randomly selected. The sample of size $n = 1,000$ is much larger than 30, so the normality condition is reasonable. Therefore, both of the necessary conditions have been satisfied.

Calculate (C): Compute the test statistic with $\bar{x} = 12.7$ hours, $s = 2$ hours, and $n = 1,000$. Therefore, $t = \frac{12.7 - 12.5}{\frac{2}{\sqrt{1000}}} = 3.1623$. This is an upper-tailed test (the inequality in H_a is >), so the P-value is the area under the t curve with 999 degrees of freedom above $t = 3.1623$. Therefore, $P\text{-value} = P(t \geq 3.1623) = 0.0008$.

Communicate Results (C): Because the P-value of 0.0008 is smaller than the specified significance level of $\alpha = 0.05$, we reject H_0. There is convincing evidence that the mean time spent online by Canadians in a typical week is greater than 12.5 hours.

(c) The sample standard deviation of 5 hours in part (a) indicates that there is likely to be more variability in the number of hours of Internet use in the population of weekly Canadian Internet users than when the sample standard deviation was 2 hours, as in part (b). If the mean of the population of weekly Canadian Internet usage was indeed 12.5 hours, it is less likely that a sample mean of 12.7 hours would be obtained when the standard deviation was 2 hours than when the standard deviation was thought to be 5 hours. Therefore, the larger standard deviation in part (a) makes it more difficult to detect a difference, resulting in a failure to reject the null hypothesis.

12.46: By saying that listening to music reduces pain levels, the authors are telling you that the study resulted in convincing evidence that pain levels are reduced when music is being listened to. In other words, the results of the study were statistically significant. However, by saying that the magnitude of the positive effect was small, the authors are telling you that the effects were not practically significant.

Section 12.3 Additional Exercises

12.53: Using the five-step process (HMC[3]):

Hypotheses (H): We want to know if the sample data provide convincing evidence that the mean supper time of a vampire bat whose meal consists of a frog is greater than 20 minutes. The population characteristic of interest is μ, the true mean supper time of a vampire bat whose meal consists of a frog.
Hypotheses: $H_0 : \mu = 20$ minutes versus $H_a : \mu > 20$ minutes.

Method (M): Because the answers to the four key questions are (Q) hypothesis testing, (S) sample data, (T) one numerical variable, and (N) one sample, we can use a one-sample t test for a population mean. The test statistic for this test is $t = \dfrac{\bar{x} - 20}{\frac{s}{\sqrt{n}}}$, with $\text{df} = n - 1 = 12 - 1 = 11$. Use a significance level of $\alpha = 0.05$.

Check (C): We are not told anything about how the bats were sampled, or whether or not they are representative of all such bats. In addition, the sample of size $n = 12$ is smaller than 30, but without additional information, we cannot assess the normality condition. Therefore, we do not have any indication that the required conditions have been satisfied. For now, we will proceed assuming the conditions have been satisfied.

Calculate (C): Compute the test statistic with $\bar{x} = 21.9$ minutes, $s = 7.7$ minutes, and $n = 12$. Therefore, $t = \dfrac{21.9 - 20}{\frac{7.7}{\sqrt{12}}} = 0.855$. This is an upper-tailed test (the inequality in H_a is >), so the P-value is the area under the t curve with 11 degrees of freedom above $t = 0.855$. Therefore, $P\text{-value} = P(t \geq 0.855) = 0.205$.

Communicate Results (C): Because the P-value of 0.205 is larger than our chosen significance level of $\alpha = 0.05$, we fail to reject H_0. There is not convincing evidence that the mean supper time of a vampire bat whose meal consists of a frog is greater than 20 minutes.

The assumptions that must be reasonable for the one-sample t test to be appropriate are a random or representative sample of bats, and a normal population distribution of supper times. In this example, there is no indication that either of the assumptions has been satisfied.

12.55: Using the five-step process (HMC^3):

Hypotheses (H): We want to know if the sample data provide convincing evidence that the mean attention span for teenage boys is less than 5 minutes. The population characteristic of interest is μ, the true mean attention span for teenage boys. Hypotheses: $H_0 : \mu = 5$ minutes versus $H_a : \mu < 5$ minutes.

Method (M): Because the answers to the four key questions are (Q) hypothesis testing, (S) sample data, (T) one numerical variable, and (N) one sample, we can use a one-sample t test for a population mean. The test statistic for this test is $t = \dfrac{\bar{x} - 5}{\frac{s}{\sqrt{n}}}$, with $\text{df} = n - 1 = 50 - 1 = 49$. Use a significance level of $\alpha = 0.01$.

Check (C): We are told to assume a random sample of 50 teenage boys. In addition, the sample of size $n = 50$ is larger than 30. Therefore, both of the necessary conditions have been satisfied.

Calculate (C): Compute the test statistic with $\overline{x} = 4$ minutes, $s = 1.4$ minutes, and a sample of size $n = 50$. Therefore, $t = \dfrac{4-5}{\dfrac{1.4}{\sqrt{50}}} = -5.051$. This is a lower-tailed test (the inequality in H_a is <), so the P-value is the area under the t curve with 49 degrees of freedom below t = -5.051. Therefore, $P\text{-value} = P(t \le -5.501) \approx 0$.

Communicate Results (C): Because the P-value of approximately 0 is smaller than the specified significance level of $\alpha = 0.01$, we reject H_0. There is convincing evidence that the mean attention span for teenage boys is less than 5 minutes.

12.57: Using the five-step process (HMC[3]):

Hypotheses (H): We want to know if the sample data provide convincing evidence that the mean fuel efficiency under these circumstances of this particular car model is less than 30 mpg. The population characteristic of interest is μ, the true mean fuel efficiency under these circumstances for this particular car model. Hypotheses: $H_0 : \mu = 30$ mpg versus $H_a : \mu < 30$ mpg.

Method (M): Because the answers to the four key questions are (Q) hypothesis testing, (S) sample data, (T) one numerical variable, and (N) one sample, we can use a one-sample t test for a population mean. The test statistic for this test is $t = \dfrac{\overline{x} - 30}{\dfrac{s}{\sqrt{n}}}$, with $\text{df} = n - 1 = 6 - 1 = 5$. Use a significance level of $\alpha = 0.05$.

Check (C): We are not told that these cars are a representative or random sample of all cars of this model. The sample of size $n = 6$ is smaller than 30, but we are told that fuel efficiencies are normally distributed. We do not have any indication that both of the required conditions have been satisfied.

Calculate (C): Compute the test statistic with $\overline{x} = 29.333$ mpg, $s = 1.408$ mpg, and a sample of size $n = 6$. Therefore, $t = \dfrac{29.333 - 30}{\dfrac{1.408}{\sqrt{6}}} = -1.160$. This is a lower-tailed test (the inequality in H_a is <), so the P-value is the area under the t curve with 5 degrees of freedom below t = -1.160. Therefore, $P\text{-value} = P(t \le -1.160) = 0.149$.

Communicate Results (C): Because the P-value of 0.149 is larger than the specified significance level of $\alpha = 0.05$, we fail to reject H_0. There is not convincing evidence that the mean fuel efficiency under these circumstances for this particular car model is less than 30 mpg.

12.59: Using the five-step process (HMC[3]):

Hypotheses (H): We want to know if the sample data provide convincing evidence that the mean time necessary to achieve a 100°F temperature is over 15 minutes. The population characteristic of interest is μ, the true mean time to achieve a 100°F temperature. Hypotheses: $H_0: \mu = 15$ minutes versus $H_a: \mu > 15$ minutes.

Method (M): Because the answers to the four key questions are (Q) hypothesis testing, (S) sample data, (T) one numerical variable, and (N) one sample, we can use a one-sample t test for a population mean. The test statistic for this test is $t = \dfrac{\bar{x} - 15}{\dfrac{s}{\sqrt{n}}}$, with $\text{df} = n - 1 = 25 - 1 = 24$. Use a significance level of $\alpha = 0.05$.

Check (C): We are told that a random sample of 25 hot tubs was selected. The sample of size n = 25 is smaller than 30, and without the data we cannot verify that the population of heating times to 100°F is normally distributed. We do not have any indication that both of the required conditions have been satisfied.

Calculate (C): Compute the test statistic with $\bar{x} = 17.5$ minutes, $s = 2.2$ minutes, and $n = 25$. Therefore, $t = \dfrac{17.5 - 15}{\dfrac{2.2}{\sqrt{25}}} = 5.682$. This is an upper-tailed test (the inequality in H_a is >), so the P-value is the area under the t curve with 24 degrees of freedom above $t = 5.682$. Therefore, $P\text{-value} = P(t \ge 5.682) \approx 0$.

Communicate Results (C): Because the P-value of approximately 0 is smaller than the specified significance level of $\alpha = 0.05$, we reject H_0. There is convincing evidence that the mean time necessary to achieve a 100°F temperature is over 15 minutes, which casts doubt on the company's claim.

Are You Ready to Move On? Chapter 12 review exercises

12.61: (a) The $\bar{x}$ distribution is centered at $\mu_{\bar{x}} = \mu = 2$ minutes and the standard deviation of the $\bar{x}$ is $\sigma_{\bar{x}} = \frac{\sigma}{\sqrt{n}} = \frac{0.8}{\sqrt{9}} = 0.267$ minutes.

(b) In each case, the $\bar{x}$ distribution is centered at $\mu_{\bar{x}} = \mu = 2$ minutes. When $n = 20$, the standard deviation of the $\bar{x}$ is $\sigma_{\bar{x}} = \frac{\sigma}{\sqrt{n}} = \frac{0.8}{\sqrt{20}} = 0.179$ minutes.

When $n = 100$, the standard deviation of the $\bar{x}$ is

$\sigma_{\bar{x}} = \frac{\sigma}{\sqrt{n}} = \frac{0.8}{\sqrt{100}} = 0.08$ minutes. The three centers are all the same, and the larger the sample size, the smaller the standard deviation of the $\bar{x}$ distribution. Since the sampling distribution of $\bar{x}$ when $n = 100$ has the smallest standard deviation, this sample size is most likely to result in a value of $\bar{x}$ close to μ.

12.63: (a) A 90% confidence interval would have been narrower because the *t* critical value for 90% confidence is smaller than the *t* critical value for 95% confidence for the same sample size, or, equivalently, for the same degrees of freedom.

(b) The statement is not correct. The population mean, μ, is a constant, and therefore, we cannot talk about the probability that it falls within a certain interval.

(c) The statement is not correct. We can say that *on average* 95 out of every 100 samples will result in confidence intervals that will contain μ, but we cannot say that in 100 such samples, *exactly* 95 will result in confidence intervals that contain μ.

12.65: Using the five-step process (EMC^3):

Estimate (E): The mean amount of money spent per graduation gift in 2007, μ, will be estimated.

Method (M): Because the answers to the four key questions are (Q) estimation, (S) sample data, (T) one numerical variable, and (N) one sample, a one-sample t confidence interval for a population mean will be considered. A confidence level of 98% was specified.

Check (C): There are two conditions that need to be met in order for the one-sample t confidence interval to be appropriate. The sample was selected to be representative of the population of adult Americans who purchased graduation gifts in 2007. Additionally, the sample size, $n = 2{,}815$, is significantly larger than 30, so it is appropriate to use the t confidence interval. Both conditions have been satisfied.

Calculate (C): From Appendix Table 3, the t critical value is 2.33 (from the z critical value row because $df = 2{,}815 - 1 = 2{,}814 > 120$, the largest number of degrees of freedom from the table). The 98% confidence interval for μ is then:

$$\bar{x} \pm (t \text{ critical value})\frac{s}{\sqrt{n}}$$

$$55.05 \pm 2.33\frac{20}{\sqrt{2{,}815}}$$

$$55.05 \pm 2.33(0.3770)$$

$$55.05 \pm 0.8783$$

$$(54.172, 55.928)$$

Communicate Results (C):

Interpret Confidence Interval: Based on this sample, you can be 98% confident that the mean amount of money spent per graduation gift in 2007 is between \$54.172 and \$55.928.

Interpret Confidence Level: The method used to construct this interval estimate is successful in capturing the actual value of the population mean about 98% of the time.

12.67: Using the sample standard deviation of s = 70 msec as a preliminary estimate of the population standard deviation, the required sample size needed for a margin of error of less than 5 msec is $n = \left(\frac{1.96\sigma}{M}\right)^2 = \left(\frac{1.96(70)}{5}\right)^2 = 752.954$, so a sample size of 753 is needed.

12.69: Using the five-step process (HMC^3):

Hypotheses (H): We want to know if the sample data provide convincing evidence that the mean wrist extension for people using this new mouse design is greater than 20 degrees. The population characteristic of interest is μ, the true mean wrist extension for people using this new mouse design. Hypotheses: $H_0 : \mu = 20$ degrees versus $H_a : \mu > 20$ degrees.

Method (M): Because the answers to the four key questions are (Q) hypothesis testing, (S) sample data, (T) one numerical variable, and (N) one sample, we can use a one-sample t test for a population mean. The test statistic for this test is $t = \frac{\bar{x} - 20}{\frac{s}{\sqrt{n}}}$, with df $= n - 1 = 24 - 1 = 23$. Use a significance level of $\alpha = 0.05$.

Check (C): We are not told how the sample of 24 Cornell University students was selected, or if they are representative of the population of Cornell University students. In addition, the sample of size $n = 24$ is smaller than 30, so we need to use the boxplot of wrist extension angles to determine if the normality condition is reasonable.

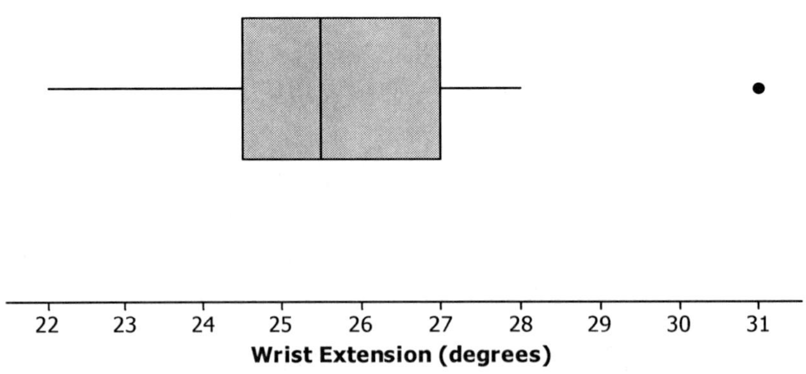

The boxplot is somewhat skewed and contains one outlier, so it is not reasonable to think that the population distribution is approximately normal. We have not satisfied both conditions, so we must proceed with caution.

Calculate (C): Compute the test statistic with $\bar{x} = 25.917$ degrees, $s = 1.954$ degrees, and a sample of size $n = 24$. Therefore, $t = \dfrac{25.917 - 20}{\dfrac{1.954}{\sqrt{24}}} = 14.835$. This is an upper-tailed test (the inequality in H_a is >), so the P-value is the area under the t curve with 23 degrees of freedom above $t = 14.835$. Therefore, $P\text{-value} = P(t \geq 14.835) \approx 0$.

Communicate Results (C): Because the P-value of approximately 0 is smaller than our chosen significance level of $\alpha = 0.05$, we reject H_0. There is convincing evidence that the mean wrist extension for people using this new mouse design is greater than 20 degrees.

To generalize the result to the population of Cornell students, we need to assume that the 24 students used in the study are representative of all students at the university. To generalize the result to the population of all university students, we need to assume that the 24 students used in the study are representative of all university students.

Chapter 13 Asking and Answering Questions about the Difference between Two Population Means

Section 13.1 Exercise Set 1

13.1: Studies 1 and 4 have samples that are independently selected. Note that in Studies 1 and 4, two separate random samples were taken, and that in Studies 2 and 3, only one random sample was taken.

13.2: Scenario 1: The appropriate test is not about a difference in population means. There is only one sample.

Scenario 2: The appropriate test is about a difference in population means.

Scenario 3: The appropriate test is not about a difference in population means. There are two samples, but the variable is categorical rather than numerical.

13.3: Using the five-step process (HMC^3):

Hypotheses (H):

We want to determine if there is convincing evidence that the mean time male students at this university spend using a computer is greater than the mean time for female students. The population characteristics of interest are μ_M, the mean time male students at this university spend using a computer, μ_F, the mean time female students at this university spend using a computer, and $\mu_M - \mu_F$, the difference in population means. Test the hypotheses $H_0 : \mu_M - \mu_F = 0$ versus $H_a : \mu_M - \mu_F > 0$.

Method (M):

Because the answers to the four key questions are (Q) hypothesis testing, (S) sample data, (T) one numerical variable, and (N) two independent samples, we will consider a two-sample t test for a difference in population means. The test statistic is: $t = \dfrac{(\bar{x}_M - \bar{x}_F) - (\mu_M - \mu_F)}{\sqrt{\dfrac{s_M^2}{n_M} + \dfrac{s_F^2}{n_F}}}$, and we will use a significance level of $\alpha = 0.05$.

Check (C):

We are told that the two samples were randomly selected and are, therefore, independent. Because both of the sample sizes are larger than 30 ($n_M = 46$ and $n_F = 38$), the large sample condition is met. The three required conditions are satisfied.

Calculations (C):

Summary statistics: $n_M = 46$, $\bar{x}_M = 45.8$ minutes, $s_M = 63.3$ minutes, $n_F = 38$, $\bar{x}_F = 39.4$ minutes, and $s_F = 57.3$ minutes

Test statistic: $t = \dfrac{(45.8 - 39.4) - 0}{\sqrt{\dfrac{(63.3)^2}{46} + \dfrac{(57.3)^2}{38}}} = 0.486$

Degrees of freedom: $V_M = \dfrac{s_M^2}{n_M} = \dfrac{(63.3)^2}{46} = 87.1063$,

$V_F = \dfrac{s_F^2}{n_F} = \dfrac{(57.3)^2}{38} = 86.4024$ and

$\text{df} = \dfrac{(V_M + V_F)^2}{\dfrac{V_M^2}{n_M - 1} + \dfrac{V_F^2}{n_F - 1}} = \dfrac{(87.1063 + 86.4024)^2}{\dfrac{(87.1063)^2}{46 - 1} + \dfrac{(86.4024)^2}{38 - 1}} = 81.2825$, truncated to $\text{df} = 81$.

This is an upper-tailed test (the inequality in H_a is >), so the P-value is the area under the t curve and to the right of 0.486 with 81 degrees of freedom. Therefore, the P-value is $P(t \geq 0.486) = 0.314$.

Communicate Results (C):

Because the P-value of 0.314 is greater than the selected significance level of $\alpha = 0.05$, we fail to reject the null hypothesis. There is not convincing evidence that the mean time male students at this university spend using a computer is greater than the mean time for female students.

13.4: Using the five-step process (HMC^3):

Hypotheses (H):

We want to determine if there is convincing evidence that the mean time spent studying for Facebook users at this university is less than the mean time spent studying by students at the university who do not use Facebook. The population characteristics of interest are μ_F , the mean time spent studying for Facebook users at this university, μ_{NF} , the mean time spent studying by students at the university who do not use Facebook, and $\mu_F - \mu_{NF}$, the difference in population means. Test the hypotheses $H_0 : \mu_F - \mu_{NF} = 0$ versus $H_a : \mu_F - \mu_{NF} < 0$.

Method (M):

Because the answers to the four key questions are (Q) hypothesis testing, (S) sample data, (T) one numerical variable, and (N) two independent samples, we will consider a two-sample *t* test for a difference in population means. The test statistic is: $t = \dfrac{(\bar{x}_F - \bar{x}_{NF}) - (\mu_F - \mu_{NF})}{\sqrt{\dfrac{s_F^2}{n_F} + \dfrac{s_{NF}^2}{n_{NF}}}}$, and we will use a significance level of $\alpha = 0.01$.

Check (C):

We are told that the samples were independent and selected to be representative of the two populations. In addition, because both of the sample sizes are larger than 30 ($n_F = 141$ and $n_{NF} = 68$), the large sample condition is met. The three required conditions are satisfied.

Calculations (C):

Summary statistics: $n_F = 141$, $\bar{x}_F = 1.47$ hours, $s_F = 0.83$ hours, $n_{NF} = 68$, $\bar{x}_{NF} = 2.76$ hours, and $s_{NF} = 0.99$ hours

Test statistic: $t = \dfrac{(1.47 - 2.76) - 0}{\sqrt{\dfrac{(0.83)^2}{141} + \dfrac{(0.99)^2}{68}}} = -9.286$

Degrees of freedom: $V_F = \frac{s_F^2}{n_F} = \frac{(0.83)^2}{141} = 0.004886$,

$V_{NF} = \frac{s_{NF}^2}{n_{NF}} = \frac{(0.99)^2}{68} = 0.014413$ and

$$\text{df} = \frac{(V_F + V_{NF})^2}{\frac{V_F^2}{n_F - 1} + \frac{V_{NF}^2}{n_{NF} - 1}} = \frac{(0.004886 + 0.014413)^2}{\frac{(0.004886)^2}{141-1} + \frac{(0.014413)^2}{68-1}} = 113.863\text{, truncated to}$$

$\text{df} = 113$.

This is a lower-tailed test (the inequality in H_a is <), so the *P*-value is the area under the *t* curve and to the left of -9.286 with 113 degrees of freedom. Therefore, the *P*-value is $P(t \le -9.286) \approx 0$.

Communicate Results (C):

Because the *P*-value of approximately 0 is less than the selected significance level of $\alpha = 0.01$, we reject the null hypothesis. There is convincing evidence that the mean time spent studying for Facebook users at this university is less than the mean time spent studying by students at the university who do not use Facebook.

13.5: Using the five-step process (HMC^3):

Hypotheses (H):

We want to determine if there is convincing evidence that the mean wait time differs for males and females. The population characteristics of interest are μ_M, the mean wait time for males, μ_F, the mean wait time for females, and $\mu_M - \mu_F$, the difference in population means. Test the hypotheses $H_0 : \mu_M - \mu_F = 0$ versus $H_a : \mu_M - \mu_F \ne 0$.

Method (M):

Because the answers to the four key questions are (Q) hypothesis testing, (S) sample data, (T) one numerical variable, and (N) two independent samples, we will consider a two-sample *t* test for a difference in population means. The test statistic is: $t=\dfrac{(\bar{x}_M-\bar{x}_F)-(\mu_M-\mu_F)}{\sqrt{\dfrac{s_M^2}{n_M}+\dfrac{s_F^2}{n_F}}}$, and we will use a significance level of $\alpha=0.05$.

Check (C):

We are told that the samples were independent and selected to be representative of the two populations. In addition, because both of the sample sizes are larger than 30 ($n_M=145$ and $n_F=141$), the large sample condition is met. The three required conditions are satisfied.

Calculations (C):

Summary statistics: $n_M=145$, $\bar{x}_M=85.2$ seconds, $s_M=50$ seconds, $n_F=141$, $\bar{x}_F=113.7$ seconds, and $s_F=75$ seconds

Test statistic: $t=\dfrac{(85.2-113.7)-0}{\sqrt{\dfrac{(50)^2}{145}+\dfrac{(75)^2}{141}}}=-3.77$

Degrees of freedom: $V_M=\dfrac{s_M^2}{n_M}=\dfrac{(50)^2}{145}=17.2414$,

$V_F=\dfrac{s_F^2}{n_F}=\dfrac{(75)^2}{141}=39.8936$ and

$\text{df}=\dfrac{(V_M+V_F)^2}{\dfrac{V_M^2}{n_M-1}+\dfrac{V_F^2}{n_F-1}}=\dfrac{(17.2414+39.8936)^2}{\dfrac{(17.2414)^2}{145-1}+\dfrac{(39.8936)^2}{141-1}}=243.029$, truncated to $\text{df}=243$.

This is a two-tailed test (the inequality in H_a is $\neq$), so the P-value is the area under the t curve and to the left of -3.77 and to the right of 3.77 with 243 degrees of freedom. Therefore, using the symmetry of the t curve, the P-value is $2\cdot P(t\le -3.77)\approx 0$.

Communicate Results (C):

Because the P-value of approximately 0 is less than the selected significance level of $\alpha = 0.05$, we reject the null hypothesis. There is convincing evidence that the mean wait time differs for males and females.

13.6: Using the five-step process (HMC^3):

Hypotheses (H):

We want to determine if there is convincing evidence that the mean braking reaction time is different for drivers talking on a cell phone and drivers who have a blood alcohol level of 0.08%. The population characteristics of interest are μ_C, the mean braking reaction time for drivers talking on a cell phone, μ_A, the mean braking reaction time for drivers who have a blood alcohol level of 0.08%, and $\mu_C - \mu_A$, the difference in population means. Test the hypotheses $H_0 : \mu_C - \mu_A = 0$ versus $H_a : \mu_C - \mu_A \neq 0$.

Method (M):

Because the answers to the four key questions are (Q) hypothesis testing, (S) sample data, (T) one numerical variable, and (N) two independent samples, we will consider a two-sample t test for a difference in population means. The test statistic is: $t = \dfrac{(\bar{x}_C - \bar{x}_A) - (\mu_C - \mu_A)}{\sqrt{\dfrac{s_C^2}{n_C} + \dfrac{s_A^2}{n_A}}}$, and we will use a significance level of $\alpha = 0.05$.

Check (C):

We are told that the samples were selected to be representative of the two populations. We don't know if the samples are independent, but for this exercise we will assume that they are. In addition, because both of the sample sizes are larger than 30 ($n_C = 40$ and $n_A = 40$), the large sample condition is met. The three required conditions are satisfied.

Calculations (C):

Summary statistics: $n_C = 40$, $\bar{x}_C = 779$ milliseconds, $s_C = 209$ milliseconds, $n_A = 40$, $\bar{x}_A = 849$ milliseconds, and $s_A = 228$ milliseconds

Test statistic: $t = \dfrac{(779-849)-0}{\sqrt{\dfrac{(209)^2}{40}+\dfrac{(228)^2}{40}}} = -1.431$

Degrees of freedom: $V_C = \dfrac{s_C^2}{n_C} = \dfrac{(209)^2}{40} = 1092.03$,

$V_A = \dfrac{s_A^2}{n_A} = \dfrac{(228)^2}{40} = 1299.6$ and

$\text{df} = \dfrac{(V_C+V_A)^2}{\dfrac{V_C^2}{n_C-1}+\dfrac{V_A^2}{n_A-1}} = \dfrac{(1092.03+1299.6)^2}{\dfrac{(1092.03)^2}{40-1}+\dfrac{(1299.6)^2}{40-1}} = 77.4168$, truncated to $\text{df} = 77$.

This is a two-tailed test (the inequality in H_a is $\neq$), so the P-value is the area under the t curve and to the left of -1.431 and to the right of 1.431 with 77 degrees of freedom. Therefore, using the symmetry of the t curve, the P-value is $2 \cdot P(t \le -1.431) = 0.156$.

Communicate Results (C):

Because the P-value of 0.156 is greater than the selected significance level of $\alpha = 0.05$, we fail to reject the null hypothesis. There is not convincing evidence that the mean braking reaction time is different for drivers talking on a cell phone and drivers who have a blood alcohol level of 0.08%.

13.7: Using the five-step process (HMC^3):

Hypotheses (H):

We want to determine if the researchers' hypothesis that the mean percentage of the time spent with the previous partner is significantly greater for genetically altered voles than for voles that did not have the gene introduced. The population characteristics of interest are μ_{GA}, the mean percentage of the time spent with the previous partner for genetically altered voles, μ_{NGA}, the mean percentage of the time spent with the previous partner for non-genetically altered voles, and $\mu_{GA} - \mu_{NGA}$, the difference in population means. Test the hypotheses $H_0 : \mu_{GA} - \mu_{NGA} = 0$ versus $H_a : \mu_{GA} - \mu_{NGA} > 0$.

Method (M):

Because the answers to the four key questions are (Q) hypothesis testing, (S) sample data, (T) one numerical variable, and (N) two independent samples, we will consider a two-sample *t* test for a difference in population means. The test statistic is: $t = \dfrac{(\bar{x}_{GA} - \bar{x}_{NGA}) - (\mu_{GA} - \mu_{NGA})}{\sqrt{\dfrac{s_{GA}^2}{n_{GA}} + \dfrac{s_{NGA}^2}{n_{NGA}}}}$,

and we will use a significance level of $\alpha = 0.05$.

Check (C):

We are not told that the samples were selected to be representative of the two populations, and we don't know if the samples are independent, but for this exercise we will assume that these conditions have been met. In addition, because both of the sample sizes are smaller than 30 ($n_{GA} = 11$ and $n_{NGA} = 20$), we need to be willing to assume that the distributions of the percent of time spent with previous partner are approximately normal for each of the two populations. Boxplots constructed using the sample data are shown below.

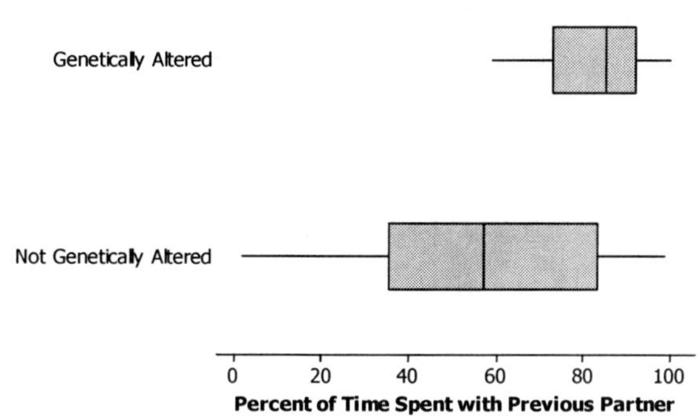

Because the boxplots are reasonably symmetric and because there are no outliers, it is reasonable to assume that the two population distributions are approximately normal. We will assume that the three required conditions are satisfied.

Calculations (C):

Summary statistics: $n_{GA} = 11$, $\bar{x}_{GA} = 82.64\,\%$, $s_{GA} = 13.10\%$, $n_{NGA} = 20$, $\bar{x}_{NGA} = 56.80\,\%$, and $s_{NGA} = 29.75\,\%$

Test statistic: $t = \dfrac{(82.64 - 56.80) - 0}{\sqrt{\dfrac{(13.10)^2}{11} + \dfrac{(29.75)^2}{20}}} = 3.340$

Degrees of freedom: $V_{GA} = \dfrac{s_{GA}^2}{n_{GA}} = \dfrac{(13.10)^2}{11} = 15.6009$,

$V_{NGA} = \dfrac{s_{NGA}^2}{n_{NGA}} = \dfrac{(29.75)^2}{20} = 44.2531$ and

$\text{df} = \dfrac{(V_{GA} + V_{NGA})^2}{\dfrac{V_{GA}^2}{n_{GA} - 1} + \dfrac{V_{NGA}^2}{n_{NGA} - 1}} = \dfrac{(15.6009 + 44.2531)^2}{\dfrac{(15.6009)^2}{11-1} + \dfrac{(44.2531)^2}{20-1}} = 28.1181$, truncated to

$\text{df} = 28$.

This is an upper-tailed test (the inequality in H_a is >), so the *P*-value is the area under the *t* curve and to the right of 3.340 with 28 degrees of freedom. Therefore, the *P*-value is $P(t \geq 3.340) = 0.0012$.

Communicate Results (C):

Because the *P*-value of 0.0012 is less than the selected significance level of $\alpha = 0.05$, we reject the null hypothesis. There is convincing evidence that the mean percentage of the time spent with the previous partner is significantly greater for genetically altered voles than for voles that did not have the gene introduced.

Section 13.1 Additional Exercises

13.15: Using the five-step process (HMC^3):

Hypotheses (H):

We want to determine if there is convincing evidence that the mean range of motion for pitchers is less than the mean for position players. The population characteristics of interest are μ_{PI}, the mean range of motion for pitchers, μ_{PP}, the mean range of motion for position players, and $\mu_{PI} - \mu_{PP}$, the difference in population means. Test the hypotheses $H_0 : \mu_{PI} - \mu_{PP} = 0$ versus $H_a : \mu_{PI} - \mu_{PP} < 0$.

Method (M):

Because the answers to the four key questions are (Q) hypothesis testing, (S) sample data, (T) one numerical variable, and (N) two independent samples, we will consider a two-sample *t* test for a difference in population means. The test statistic is: $t = \dfrac{(\bar{x}_{PI} - \bar{x}_{PP}) - (\mu_{PI} - \mu_{PP})}{\sqrt{\dfrac{s_{PI}^2}{n_{PI}} + \dfrac{s_{PP}^2}{n_{PP}}}}$, and we will use a significance level of $\alpha = 0.05$.

Check (C):

We are told that the samples were independent and selected to be representative of the two populations. In addition, because both of the sample sizes are larger than 30 ($n_{PI} = 40$ and $n_{PP} = 40$), the large sample condition is met. The three required conditions are satisfied.

Calculations (C):

Summary statistics: $n_{PI} = 40$, $\bar{x}_{PI} = 75.6$ degrees, $s_{PI} = 5.9$ degrees, $n_{PP} = 40$, $\bar{x}_{PP} = 79.6$ degrees, and $s_{PP} = 7.6$ degrees

Test statistic: $t = \dfrac{(75.6 - 79.6) - 0}{\sqrt{\dfrac{(5.9)^2}{40} + \dfrac{(7.6)^2}{40}}} = -2.629$

Degrees of freedom: $V_{PI} = \frac{s_{PI}^2}{n_{PI}} = \frac{(5.9)^2}{40} = 0.87025$,

$$V_{PP} = \frac{s_{PP}^2}{n_{PP}} = \frac{(7.6)^2}{40} = 1.444 \text{ and}$$

$$\text{df} = \frac{(V_{PI} + V_{PP})^2}{\frac{V_{PI}^2}{n_{PI}-1} + \frac{V_{PP}^2}{n_{PP}-1}} = \frac{(0.87025 + 1.444)^2}{\frac{(0.87025)^2}{40-1} + \frac{(1.444)^2}{40-1}} = 73.4834 \text{, truncated to}$$

$\text{df} = 73$.

This is a lower-tailed test (the inequality in H_a is <), so the P-value is the area under the t curve and to the left of -2.629 with 73 degrees of freedom. Therefore, the P-value is $P(t \le -2.629) = 0.005$.

Communicate Results (C):

Because the P-value of 0.005 is less than the selected significance level of $\alpha = 0.05$, we reject the null hypothesis. There is convincing evidence that the mean range of motion for pitchers is less than the mean for position players.

13.17: Using the five-step process (HMC^3):

Hypotheses (H):

We want to determine if there is convincing evidence that the mean brain volume of children with ADHD is smaller than the mean for children without ADHD. The population characteristics of interest are μ_W, the mean brain volume of children with ADHD, $\mu_{W/O}$, the mean brain volume of children without ADHD, and $\mu_W - \mu_{W/O}$, the difference in population means. Test the hypotheses $H_0 : \mu_W - \mu_{W/O} = 0$ versus $H_a : \mu_W - \mu_{W/O} < 0$.

Method (M):

Because the answers to the four key questions are (Q) hypothesis testing, (S) sample data, (T) one numerical variable, and (N) two independent samples, we will consider a two-sample t test for a difference in population means. The test statistic is: $t=\frac{(\bar{x}_W-\bar{x}_{W/O})-(\mu_W-\mu_{W/O})}{\sqrt{\frac{s_W^2}{n_W}+\frac{s_{W/O}^2}{n_{W/O}}}}$, and we will use a significance level of $\alpha=0.05$.

Check (C):

We are told that the samples were independent and selected to be representative of the two populations. In addition, because both of the sample sizes are larger than 30 ($n_W=152$ and $n_{W/O}=139$), the large sample condition is met. The three required conditions are satisfied.

Calculations (C):

Summary statistics: $n_W=152$, $\bar{x}_W=1059.4$ milliliters, $s_W=117.5$ milliliters, $n_{W/O}=139$, $\bar{x}_{W/O}=1104.5$ milliliters, and $s_{W/O}=111.3$ milliliters

Test statistic: $t=\frac{(1059.4-1104.5)-0}{\sqrt{\frac{(117.5)^2}{152}+\frac{(111.3)^2}{139}}}=-3.362$

Degrees of freedom: $V_W=\frac{s_W^2}{n_W}=\frac{(117.5)^2}{152}=90.8306$,

$V_{W/O}=\frac{s_{W/O}^2}{n_{W/O}}=\frac{(111.3)^2}{139}=89.1201$ and

$\text{df}=\frac{(V_W+V_{W/O})^2}{\frac{V_W^2}{n_W-1}+\frac{V_{W/O}^2}{n_{W/O}-1}}=\frac{(90.8306+89.1201)^2}{\frac{(90.8306)^2}{152-1}+\frac{(89.1201)^2}{139-1}}=288.636$, truncated to $\text{df}=288$.

This is a lower-tailed test (the inequality in H_a is <), so the P-value is the area under the t curve and to the left of -3.362 with 288 degrees of freedom. Therefore, the P-value is $P(t\le-3.362)=0.0004$.

Communicate Results (C):

Because the P-value of 0.0004 is less than the selected significance level of $\alpha = 0.05$, we reject the null hypothesis. There is convincing evidence that the mean brain volume of children with ADHD is smaller than the mean for children without ADHD.

13.19: (a) The assumption that must be made about the distributions of electronic media use times is that the distributions are approximately normal.

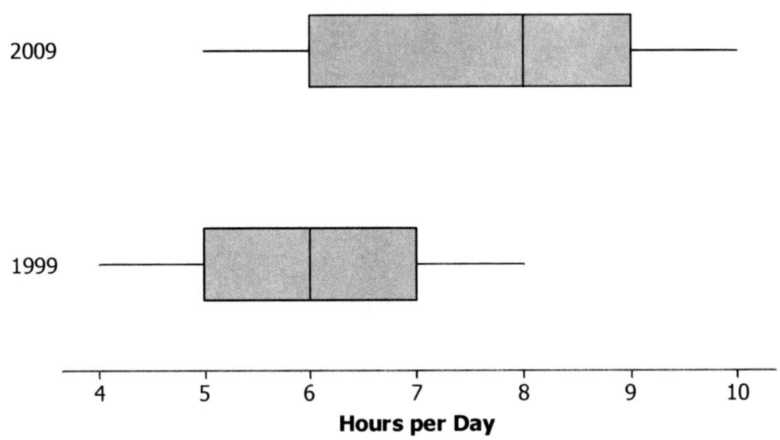

Since the boxplots are roughly symmetric and since there are no outliers in either sample, the assumption of normality is justified, and it is reasonable to carry out a two-sample t test.

(b) Using the five-step process (HMC^3):

Hypotheses (H):

We want to determine if there is convincing evidence that the mean number of hours per day spent using electronic media was greater in 2009 than in 1999. The population characteristics of interest are μ_{2009}, the mean number of hours per day spent using electronic media in 2009, μ_{1999}, the mean number of hours per day spent using electronic media in 1999, and $\mu_{2009} - \mu_{1999}$, the difference in population means. Test the hypotheses $H_0 : \mu_{2009} - \mu_{1999} = 0$ versus $H_a : \mu_{2009} - \mu_{1999} > 0$.

Method (M):

Because the answers to the four key questions are (Q) hypothesis testing, (S) sample data, (T) one numerical variable, and (N) two independent samples, we will consider a two-sample t test for a difference in population means. The test statistic is: $t = \dfrac{(\bar{x}_{2009} - \bar{x}_{1999}) - (\mu_{2009} - \mu_{1999})}{\sqrt{\dfrac{s_{2009}^2}{n_{2009}} + \dfrac{s_{1999}^2}{n_{1999}}}}$, and we will use a significance level of $\alpha = 0.01$.

Check (C):

We are told that the samples were selected to be representative of the two populations, and, given that the samples were selected 10 years apart, we will assume that the samples are independent. In addition, because both of the sample sizes are smaller than 30 ($n_{2009} = 15$ and $n_{1999} = 15$), we need to be willing to assume that the distributions are approximately normal for each of the two populations. The boxplots constructed in part (a) using the sample data are roughly symmetric and since there are no outliers in either sample, the assumption of normality is justified. Therefore, the three conditions have been satisfied and it is reasonable to carry out a two-sample t test.

Calculations (C):

Summary statistics: $n_{2009} = 15$, $\bar{x}_{2009} = 7.6$ hours, $s_{2009} = 1.595$ hours, $n_{1999} = 15$, $\bar{x}_{1999} = 5.933$ hours, and $s_{1999} = 1.100$ hours

Test statistic: $t = \dfrac{(7.600 - 5.933) - 0}{\sqrt{\dfrac{(1.595)^2}{15} + \dfrac{(1.100)^2}{15}}} = 3.332$

Degrees of freedom: $V_{2009} = \dfrac{s_{2009}^2}{n_{2009}} = \dfrac{(1.595)^2}{15} = 0.169602$,

$V_{1999} = \dfrac{s_{1999}^2}{n_{1999}} = \dfrac{(1.100)^2}{15} = 0.080667$ and

$$\text{df} = \frac{\left(V_{2009} + V_{1999}\right)^2}{\frac{V_{2009}^2}{n_{2009}-1} + \frac{V_{1999}^2}{n_{1999}-1}} = \frac{\left(0.169602 + 0.080667\right)^2}{\frac{\left(0.169602\right)^2}{15-1} + \frac{\left(0.080667\right)^2}{15-1}} = 24.8606,$$

truncated to $\text{df} = 24$.

This is an upper-tailed test (the inequality in H_a is >), so the *P*-value is the area under the *t* curve and to the right of 3.332 with 24 degrees of freedom. Therefore, the *P*-value is $P(t \geq 3.332) = 0.001$.

Communicate Results (C):

Because the *P*-value of 0.001 is less than the selected significance level of $\alpha = 0.01$, we reject the null hypothesis. There is convincing evidence that the mean number of hours per day spent using electronic media was greater in 2009 than in 1999.

13.21: (a) μ_1 = mean payment for claims not involving errors; μ_2 = mean payment for claims involving errors; H_0: $\mu_1 - \mu_2 = 0$ versus H_a: $\mu_1 - \mu_2 < 0$.

(b) Answer: (ii) 2.65. Since the samples are large, a *t* distribution with a large number of degrees of freedom is used, which can be approximated with the standard normal distribution. $P(z > 2.65) = 0.004$, which is the *P*-value given. None of the other possible values of *t* gives the correct *P*-value.

13.23: (a) Using the five-step process (HMC^3):

Hypotheses (H):

We want to determine if there is convincing evidence that the mean percentage of time spent playing with the police car is greater for male monkeys than for female monkeys. The population characteristics of interest are μ_M, the mean percentage of time male monkeys spend playing with the police car, μ_F, the mean percentage of time female monkeys spend playing with the police car, and $\mu_M - \mu_F$, the difference in population means. Test the hypotheses $H_0: \mu_M - \mu_F = 0$ versus $H_a: \mu_M - \mu_F > 0$.

Method (M):

Because the answers to the four key questions are (Q) hypothesis testing, (S) sample data, (T) one numerical variable, and (N) two independent samples, we will consider a two-sample *t* test for a difference in population means. The test statistic is: $t = \dfrac{(\bar{x}_M - \bar{x}_F) - (\mu_M - \mu_F)}{\sqrt{\dfrac{s_M^2}{n_M} + \dfrac{s_F^2}{n_F}}}$, and we will use a significance level of $\alpha = 0.05$.

Check (C):

We are told that the samples were independent and selected to be representative of the two populations. In addition, because both of the sample sizes are larger than 30 ($n_M = 44$ and $n_F = 44$), the large sample condition is met. The three required conditions are satisfied.

Calculations (C):

Summary statistics: $n_M = 44$, $\bar{x}_M = 18$ percent, $s_M = 5$ percent, $n_F = 44$, $\bar{x}_F = 8$ percent, and $s_F = 4$ percent

Test statistic: $t = \dfrac{(18-8)-0}{\sqrt{\dfrac{(5)^2}{44} + \dfrac{(4)^2}{44}}} = 10.3594$

Degrees of freedom: $V_M = \dfrac{s_M^2}{n_M} = \dfrac{(5)^2}{44} = 0.568182$,

$V_F = \dfrac{s_F^2}{n_F} = \dfrac{(4)^2}{44} = 0.363636$ and

$\text{df} = \dfrac{(V_M + V_F)^2}{\dfrac{V_M^2}{n_M - 1} + \dfrac{V_F^2}{n_F - 1}} = \dfrac{(0.568182 + 0.363636)^2}{\dfrac{(0.568182)^2}{44-1} + \dfrac{(0.363636)^2}{44-1}} = 82.0465$, truncated to $\text{df} = 82$.

This is an upper-tailed test (the inequality in H_a is >), so the P-value is the area under the t curve and to the right of 10.3594 with 82 degrees of freedom. Therefore, the P-value is $P(t \geq 10.3594) \approx 0$.

Communicate Results (C):

Because the P-value of approximately 0 is less than the selected significance level of $\alpha = 0.05$, we reject the null hypothesis. There is convincing evidence that the mean percentage of time spent playing with the police car is greater for male monkeys than for female monkeys.

(b) Using the five-step process (HMC^3):

Hypotheses (H):

We want to determine if there is convincing evidence that the mean percentage of time spent playing with the doll is greater for female monkeys than for male monkeys. The population characteristics of interest are μ_M, the mean percentage of time male monkeys spend playing with the doll, μ_F, the mean percentage of time female monkeys spend playing with the doll, and $\mu_M - \mu_F$, the difference in population means. Test the hypotheses $H_0 : \mu_M - \mu_F = 0$ versus $H_a : \mu_M - \mu_F < 0$.

Method (M):

Because the answers to the four key questions are (Q) hypothesis testing, (S) sample data, (T) one numerical variable, and (N) two independent samples, we will consider a two-sample t test for a difference in population means. The test statistic is: $t = \dfrac{(\bar{x}_M - \bar{x}_F) - (\mu_M - \mu_F)}{\sqrt{\dfrac{s_M^2}{n_M} + \dfrac{s_F^2}{n_F}}}$, and we will use a significance level of $\alpha = 0.05$.

Check (C):

We are told that the samples were independent and selected to be representative of the two populations. In addition, because both of the sample sizes are larger than 30 ($n_M = 44$ and $n_F = 44$), the large sample condition is met. The three required conditions are satisfied.

Calculations (C):

Summary statistics: $n_M = 44$, $\bar{x}_M = 9$ percent, $s_M = 2$ percent, $n_F = 44$, $\bar{x}_F = 20$ percent, and $s_F = 4$ percent

Test statistic: $t = \dfrac{(9-20)-0}{\sqrt{\dfrac{(2)^2}{44}+\dfrac{(4)^2}{44}}} = -16.3156$

Degrees of freedom: $V_M = \dfrac{s_M^2}{n_M} = \dfrac{(2)^2}{44} = 0.090909$,

$V_F = \dfrac{s_F^2}{n_F} = \dfrac{(4)^2}{44} = 0.363636$ and

$\text{df} = \dfrac{(V_M + V_F)^2}{\dfrac{V_M^2}{n_M - 1} + \dfrac{V_F^2}{n_F - 1}} = \dfrac{(0.090909 + 0.363636)^2}{\dfrac{(0.090909)^2}{44-1} + \dfrac{(0.363636)^2}{44-1}} = 63.2353$, truncated to $\text{df} = 63$.

This is a lower-tailed test (the inequality in H_a is <), so the P-value is the area under the t curve and to the left of -16.3156 with 63 degrees of freedom. Therefore, the P-value is $P(t \le -16.3156) \approx 0$.

Communicate Results (C):

Because the P-value of approximately 0 is less than the selected significance level of $\alpha = 0.05$, we reject the null hypothesis. There is convincing evidence that the mean percentage of time spent playing with the doll is greater for female monkeys than for male monkeys.

(c) Using the five-step process (HMC^3):

Hypotheses (H):

We want to determine if there is convincing evidence that the mean percentage of time spent playing with the furry dog is not the same for male and female monkeys. The population characteristics of interest are μ_M, the mean percentage of time male monkeys spend playing with the furry dog, μ_F, the mean percentage of time female monkeys spend playing with the furry dog, and $\mu_M - \mu_F$, the difference in population means. Test the hypotheses $H_0: \mu_M - \mu_F = 0$ versus $H_a: \mu_M - \mu_F \neq 0$.

Method (M):

Because the answers to the four key questions are (Q) hypothesis testing, (S) sample data, (T) one numerical variable, and (N) two independent samples, we will consider a two-sample *t* test for a difference in population means. The test statistic is: $t = \frac{(\bar{x}_M - \bar{x}_F) - (\mu_M - \mu_F)}{\sqrt{\frac{s_M^2}{n_M} + \frac{s_F^2}{n_F}}}$, and we will use a significance level of $\alpha = 0.05$.

Check (C):

We are told that the samples were independent and selected to be representative of the two populations. In addition, because both of the sample sizes are larger than 30 ($n_M = 44$ and $n_F = 44$), the large sample condition is met. The three required conditions are satisfied.

Calculations (C):

Summary statistics: $n_M = 44$, $\bar{x}_M = 25$ percent, $s_M = 5$ percent, $n_F = 44$, $\bar{x}_F = 20$ percent, and $s_F = 5$ percent

Test statistic: $t = \frac{(25-20)-0}{\sqrt{\frac{(5)^2}{44} + \frac{(5)^2}{44}}} = 4.690$

Degrees of freedom: $V_M = \frac{s_M^2}{n_M} = \frac{(5)^2}{44} = 0.568182$,

$V_F = \frac{s_F^2}{n_F} = \frac{(5)^2}{44} = 0.568182$ and

$$\text{df} = \frac{(V_M + V_F)^2}{\frac{V_M^2}{n_M - 1} + \frac{V_F^2}{n_F - 1}} = \frac{(0.568182 + 0.568182)^2}{\frac{(0.568182)^2}{44-1} + \frac{(0.568182)^2}{44-1}} = 86.$$

This is a two-tailed test (the inequality in H_a is $\neq$), so the P-value is the area under the t curve and to the left of –4.690 and to the right of 4.690 with 86 degrees of freedom. Therefore, using the symmetry of the t curve, the P-value is $2 \cdot P(t \leq -4.690) \approx 0$.

Communicate Results (C):

Because the P-value of approximately 0 is less than the selected significance level of $\alpha = 0.05$, we reject the null hypothesis. There is convincing evidence that the mean percentage of the time spent playing with the furry dog is different for males and females.

(d) The results do seem to provide convincing evidence of a gender basis in the monkeys' choices of how much time to spend playing with each toy, with the male monkeys spending significantly more time with the "masculine" toy than the female monkeys, and with the female monkeys spending significantly more time with the "feminine" toy than the male monkeys. However, the data also provide convincing evidence of a difference between male and female monkeys in the time they choose to spend playing with a "neutral" toy. It is possible that it was some attribute other than masculinity/femininity in the toys that was attracting the different genders of monkey in different ways.

(e) The given mean time playing with the police car and mean time playing with the doll for female monkeys are sample means for the same sample of female monkeys. The two-sample t test can only be performed when there are two independent random samples.

13.25: (a) Using the five-step process (HMC^3):

Hypotheses (H):

We want to determine if there is convincing evidence that the mean "appropriateness" score assigned to wearing a hat in class differs for students and faculty. The population characteristics of interest are μ_S, the student mean "appropriateness" score assigned to wearing a hat, μ_F, the faculty mean "appropriateness" score assigned to wearing a hat, and $\mu_S - \mu_F$, the difference in population means. Test the hypotheses $H_0: \mu_S - \mu_F = 0$ versus $H_a: \mu_S - \mu_F \neq 0$.

Method (M):

Because the answers to the four key questions are (Q) hypothesis testing, (S) sample data, (T) one numerical variable, and (N) two independent samples, we will consider a two-sample t test for a difference in population means. The test statistic is: $t = \frac{(\bar{x}_S - \bar{x}_F) - (\mu_S - \mu_F)}{\sqrt{\frac{s_S^2}{n_S} + \frac{s_F^2}{n_F}}}$, and we will use a significance level of $\alpha = 0.05$.

Check (C):

We are told that the samples were independent and randomly selected from the two populations. Because both of the sample sizes are larger than 30 ($n_S = 173$ and $n_F = 98$), the large sample condition is met. The three required conditions are satisfied.

Calculations (C):

Summary statistics: $n_S = 173$, $\bar{x}_S = 2.80$, $s_S = 1$, $n_F = 98$, $\bar{x}_F = 3.63$, and $s_F = 1$

Test statistic: $t = \frac{(2.80 - 3.63) - 0}{\sqrt{\frac{(1)^2}{173} + \frac{(1)^2}{98}}} = -6.565$

Degrees of freedom: $V_S = \frac{s_S^2}{n_S} = \frac{(1)^2}{173} = 0.00578$,

$V_F = \frac{s_F^2}{n_F} = \frac{(1)^2}{98} = 0.010204$ and

$$\text{df} = \frac{(V_S + V_F)^2}{\frac{V_S^2}{n_S - 1} + \frac{V_F^2}{n_F - 1}} = \frac{(0.00578 + 0.010204)^2}{\frac{(0.00578)^2}{173 - 1} + \frac{(0.010204)^2}{98 - 1}} = 201.544$$, truncated to $\text{df} = 201$.

This is a two-tailed test (the inequality in H_a is $\neq$), so the P-value is the area under the t curve and to the left of –6.565 and to the right of 6.565

with 201 degrees of freedom. Therefore, using the symmetry of the t curve, the P-value is $2 \cdot P(t \leq -6.565) \approx 0$.

Communicate Results (C):

Because the P-value of approximately 0 is less than the selected significance level of $\alpha = 0.05$, we reject the null hypothesis. There is convincing evidence that the mean appropriateness score assigned to wearing a hat in class is different for students and faculty.

(b) Using the five-step process (HMC^3):

Hypotheses (H):

We want to determine if there is convincing evidence that the mean "appropriateness" score assigned to addressing an instructor by his or her first name is higher for students than for faculty. The population characteristics of interest are μ_S, the student mean "appropriateness" score assigned to addressing an instructor by his or her first name, μ_F, the faculty mean "appropriateness" score assigned to addressing an instructor by his or her first name, and $\mu_S - \mu_F$, the difference in population means. Test the hypotheses $H_0 : \mu_S - \mu_F = 0$ versus $H_a : \mu_S - \mu_F > 0$.

Method (M):

Because the answers to the four key questions are (Q) hypothesis testing, (S) sample data, (T) one numerical variable, and (N) two independent samples, we will consider a two-sample t test for a difference in population means. The test statistic is: $t = \dfrac{(\bar{x}_S - \bar{x}_F) - (\mu_S - \mu_F)}{\sqrt{\dfrac{s_S^2}{n_S} + \dfrac{s_F^2}{n_F}}}$, and we will use a significance level of $\alpha = 0.05$.

Check (C):

We are told that the samples were independent and randomly selected from the two populations. Because both of the sample sizes are larger than 30 ($n_S = 173$ and $n_F = 98$), the large sample condition is met. The three required conditions are satisfied.

Calculations (C):

Summary statistics: $n_S = 173$, $\bar{x}_S = 2.90$, $s_S = 1$, $n_F = 98$, $\bar{x}_F = 2.11$, and $s_F = 1$

Test statistic: $t = \dfrac{(2.90 - 2.11) - 0}{\sqrt{\dfrac{(1)^2}{173} + \dfrac{(1)^2}{98}}} = 6.249$

Degrees of freedom: $V_S = \dfrac{s_S^2}{n_S} = \dfrac{(1)^2}{173} = 0.00578$,

$V_F = \dfrac{s_F^2}{n_F} = \dfrac{(1)^2}{98} = 0.010204$ and

$\text{df} = \dfrac{(V_S + V_F)^2}{\dfrac{V_S^2}{n_S - 1} + \dfrac{V_F^2}{n_F - 1}} = \dfrac{(0.00578 + 0.010204)^2}{\dfrac{(0.00578)^2}{173 - 1} + \dfrac{(0.010204)^2}{98 - 1}} = 201.544$, truncated to $\text{df} = 201$.

This is an upper-tailed test (the inequality in H_a is >), so the P-value is the area under the t curve and to the right of 6.249 with 201 degrees of freedom. Therefore, the P-value is $P(t \geq 6.249) \approx 0$.

Communicate Results (C):

Because the P-value of approximately 0 is less than the selected significance level of $\alpha = 0.05$, we reject the null hypothesis. There is convincing evidence that the mean appropriateness score assigned to addressing an instructor by his or her first name is higher for students than for faculty.

13.27: Using the five-step process (HMC^3):

Hypotheses (H):

We want to determine if there is convincing evidence that the mean number of goals scored by the Oilers is higher for games when Gretzky plays. The population characteristics of interest are μ_G, the mean number of goals scored when Gretzky plays, $\mu_{W/O}$, the mean number of goals scored when Gretzky does not play, and $\mu_G - \mu_{W/O}$, the difference in population means. Test the hypotheses $H_0: \mu_G - \mu_{W/O} = 0$ versus $H_a: \mu_G - \mu_{W/O} > 0$.

Method (M):

Because the answers to the four key questions are (Q) hypothesis testing, (S) sample data, (T) one numerical variable, and (N) two independent samples, we will consider a two-sample *t* test for a difference in population means. The test statistic is: $t = \dfrac{(\bar{x}_G - \bar{x}_{W/O}) - (\mu_G - \mu_{W/O})}{\sqrt{\dfrac{s_G^2}{n_G} + \dfrac{s_{W/O}^2}{n_{W/O}}}}$, and we will use a significance level of $\alpha = 0.01$.

Check (C):

We are told to assume that the samples were independent and randomly selected from the two populations. One of the sample sizes is larger than 30 ($n_G = 41$) and the other is smaller than 30 ($n_{W/O} = 17$), and so the large sample condition has not been met. Without additional information, we cannot verify the normality condition, but we will proceed as if all conditions have been satisfied.

Calculations (C):

Summary statistics: $n_G = 41$, $\bar{x}_G = 4.73$, $s_G = 1.29$, $n_{W/O} = 17$, $\bar{x}_{W/O} = 3.88$, and $s_{W/O} = 1.18$

Test statistic: $t = \dfrac{(4.73 - 3.88) - 0}{\sqrt{\dfrac{(1.29)^2}{41} + \dfrac{(1.18)^2}{17}}} = 2.429$

Degrees of freedom: $V_G = \dfrac{s_G^2}{n_G} = \dfrac{(1.29)^2}{41} = 0.040588$,

$V_{W/O} = \dfrac{s_{W/O}^2}{n_{W/O}} = \dfrac{(1.18)^2}{17} = 0.081906$ and

$$\text{df} = \frac{\left(V_G + V_{W/O}\right)^2}{\frac{V_G^2}{n_G - 1} + \frac{V_{W/O}^2}{n_{W/O} - 1}} = \frac{\left(0.040588 + 0.081906\right)^2}{\frac{\left(0.040588\right)^2}{41-1} + \frac{\left(0.081906\right)^2}{17-1}} = 32.5857 \text{, truncated}$$

to $\text{df} = 32$.

This is an upper-tailed test (the inequality in H_a is >), so the P-value is the area under the t curve and to the right of 2.429 with 32 degrees of freedom. Therefore, the P-value is $P(t \geq 2.429) = 0.0105$.

Communicate Results (C):

Because the P-value of 0.0105 is greater than the selected significance level of $\alpha = 0.01$, we fail to reject the null hypothesis. There is not convincing evidence that the mean number of goals scored was higher for games in which Gretzky played than for games in which he did not play.

Section 13.2 Exercise Set 1

13.28: Using the five-step process (HMC^3):

Hypotheses (H):

We want to determine if there is convincing evidence that the mean reading differs for the Mini-Wright meter and the Wright meter. The population characteristics of interest are μ_{MW}, the mean Mini-Wright meter reading, μ_W, the mean Wright meter reading, and $\mu_d = \mu_{MW} - \mu_W$, the difference in the paired means. Test the hypotheses $H_0 : \mu_d = 0$ versus $H_a : \mu_d \neq 0$.

Method (M):

Because the answers to the four key questions are (Q) hypothesis testing, (S) sample data, (T) one numerical variable, and (N) two paired samples, we will consider the paired-samples t test as a potential method. The test statistic is $t = \dfrac{\bar{x}_d - \mu_0}{\frac{s_d}{\sqrt{n}}}$, and we will use a significance level of $\alpha = 0.01$.

Check (C):

We are told to assume that it is reasonable to consider the 17 Wright meter measurements and the 17 Mini-Wright meter measurements as random samples from their respective populations. Because the sample size is small ($n=17$), we must be willing to assume that the distribution of differences for the population of Mini-Wright and Wright meter readings is at least approximately normal. The following boxplot of the seventeen sample differences is fairly symmetric and there are no outliers, so it is not unreasonable to think that the population distribution could be at least approximately normal. These observations suggest that it is appropriate to use the paired-sample t test.

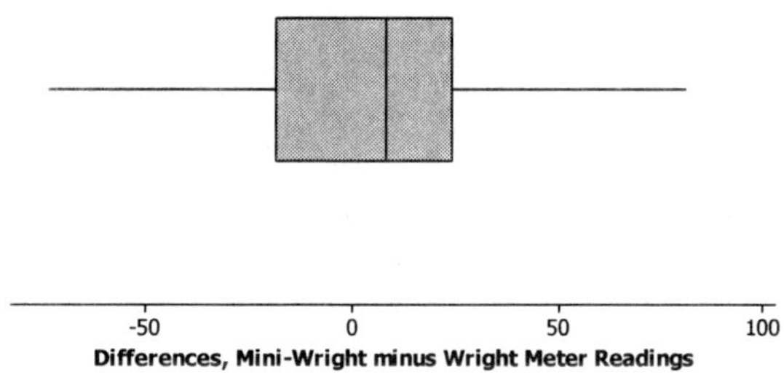

Calculations (C):

Summary statistics: $n=17$, $\bar{x}_d=2.12$ and $s_d=38.77$

Test statistic: $t=\dfrac{2.12-0}{\dfrac{38.77}{\sqrt{17}}}=0.225$

Degrees of freedom when the null hypothesis is true is $\text{df}=n-1=17-1=16$.

This is a two-tailed test (the inequality in H_a is $\neq$), so the P-value is the total area under the t curve and to the left of –0.225 and to the right of 0.225 with 16 degrees of freedom. Therefore, using the symmetry of the t curve, the P-value is $2\cdot P(t\leq -0.225)=0.825$.

Communicate Results (C):

Because the P-value of 0.825 is greater than the selected significance level of $\alpha = 0.01$, we fail to reject the null hypothesis. There is not convincing evidence that the mean reading differs for the Mini-Wright meter and the Wright meter.

13.29: Using the five-step process (HMC[3]):

Hypotheses (H):

We want to determine if there is convincing evidence that the mean time to exhaustion is greater after chocolate milk than after a carbohydrate replacement drink. The population characteristics of interest are μ_{CM}, the mean time to exhaustion after chocolate milk, μ_{CR}, the mean time to exhaustion after a carbohydrate replacement drink, and $\mu_d = \mu_{CM} - \mu_{CR}$, the difference in the paired means. Test the hypotheses $H_0 : \mu_d = 0$ versus $H_a : \mu_d > 0$.

Method (M):

Because the answers to the four key questions are (Q) hypothesis testing, (S) sample data, (T) one numerical variable, and (N) two paired samples, we will consider the paired-samples t test as a potential method. The test statistic is $t = \frac{\bar{x}_d - \mu_0}{\frac{s_d}{\sqrt{n}}}$, and we will use a significance level of $\alpha = 0.05$.

Check (C):

The exercise gives no indication of how the cyclists were selected, or whether they are representative of the population of all cyclists. Because the sample size is small ($n = 9$), we must be willing to assume that the distribution of differences in the population of time to exhaustion for the chocolate milk and carbohydrate replacement drink is at least approximately normal. The following boxplot of the nine sample differences is fairly symmetric and there are no outliers, so it is not unreasonable to think that the population distribution could be at least approximately normal. These observations suggest that it is appropriate to use the paired-sample t test.

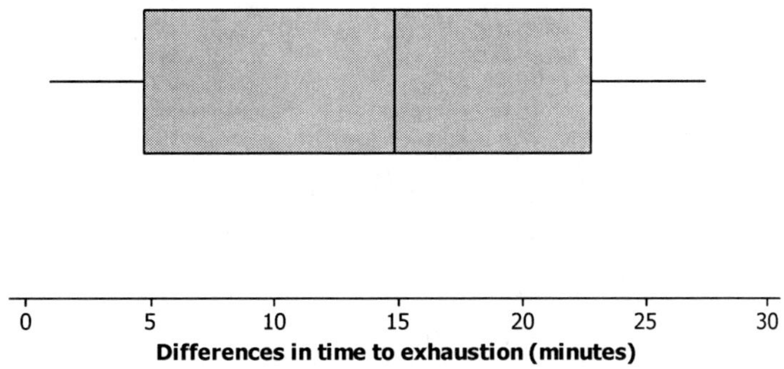

Calculations (C):

Summary statistics: $n = 9$, $\bar{x}_d = 14.08$ and $s_d = 9.47$

Test statistic: $t = \dfrac{14.08 - 0}{\frac{9.47}{\sqrt{9}}} = 4.46$

Degrees of freedom when the null hypothesis is true is $\text{df} = n - 1 = 9 - 1 = 8$.

This is an upper-tailed test (the inequality in H_a is >), so the *P*-value is the total area under the *t* curve and to the right of 4.46 with 8 degrees of freedom. Therefore, the *P*-value is $P(t \geq 4.46) = 0.001$.

Communicate Results (C):

Because the *P*-value of 0.001 is less than the selected significance level of $\alpha = 0.05$, we reject the null hypothesis. There is convincing evidence that the mean time to exhaustion is greater after chocolate milk than after carbohydrate replacement drink.

13.30: Using the five-step process (HMC^3):

Hypotheses (H):

We want to determine if there is convincing evidence that the mean MPF at brain location 1 is higher after diesel exposure than before diesel exposure. The population characteristics of interest are μ_{1B}, the mean MPF at brain location 1 before diesel exposure, μ_{1A}, the mean MPF at brain location 1 after diesel exposure, and $\mu_d = \mu_{1B} - \mu_{1A}$, the difference in the paired means. Test the hypotheses $H_0: \mu_d = 0$ versus $H_a: \mu_d < 0$.

Method (M):

Because the answers to the four key questions are (Q) hypothesis testing, (S) sample data, (T) one numerical variable, and (N) two paired samples, we will consider the paired-samples *t* test as a potential method. The test statistic is $t = \dfrac{\bar{x}_d - \mu_0}{\dfrac{s_d}{\sqrt{n}}}$, and we will use a significance level of $\alpha = 0.05$.

Check (C):

We are told to assume that it is reasonable to regard the sample of 10 men as representative of healthy adult males. Because the sample size is small ($n = 10$), we must be willing to assume that the distribution of differences in MPF before and after diesel exposure is at least approximately normal. The following boxplot of the ten sample differences is fairly symmetric and there are no outliers, so it is not unreasonable to think that the population distribution could be at least approximately normal. These observations suggest that it is appropriate to use the paired-sample *t* test.

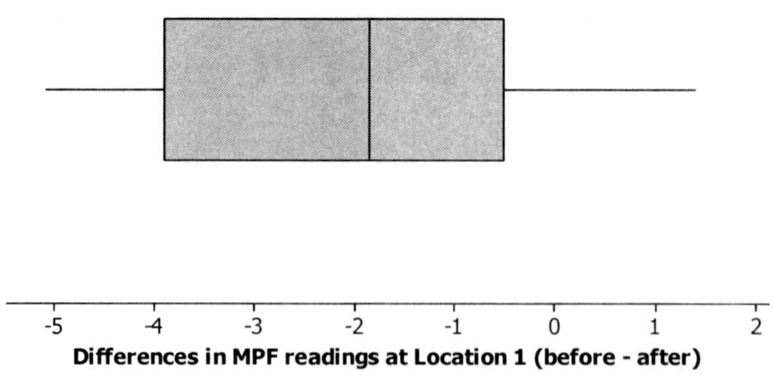

Calculations (C):

Summary statistics: $n = 10$, $\bar{x}_d = -1.930$ and $s_d = 1.965$

Test statistic: $t = \dfrac{-1.930 - 0}{\dfrac{1.965}{\sqrt{10}}} = -3.11$

Degrees of freedom when the null hypothesis is true is
$\text{df} = n - 1 = 10 - 1 = 9$.

This is a lower-tailed test (the inequality in H_a is <), so the P-value is the total area under the t curve and to the left of –3.11 with 9 degrees of freedom. Therefore, the P-value is $P(t \le -3.11) = 0.006$.

Communicate Results (C):

Because the P-value of 0.006 is smaller than the selected significance level of $\alpha = 0.05$, we reject the null hypothesis. There is convincing evidence that the mean MPF at brain location 1 is higher after diesel exposure than before diesel exposure.

13.31: Using the five-step process (HMC^3):

Hypotheses (H):

We want to determine if there is convincing evidence that the mean MPF at brain location 2 is higher after diesel exposure than before diesel exposure. The population characteristics of interest are μ_{2B}, the mean MPF at brain location 2 before diesel exposure, μ_{2A}, the mean MPF at brain location 2 after diesel exposure, and $\mu_d = \mu_{2B} - \mu_{2A}$, the difference in the paired means. Test the hypotheses $H_0 : \mu_d = 0$ versus $H_a : \mu_d < 0$.

Method (M):

Because the answers to the four key questions are (Q) hypothesis testing, (S) sample data, (T) one numerical variable, and (N) two paired samples,

we will consider the paired-samples t test as a potential method. The test statistic is $t = \dfrac{\bar{x}_d - \mu_0}{\dfrac{s_d}{\sqrt{n}}}$, and we will use a significance level of $\alpha = 0.05$.

Check (C):

We are told to assume that it is reasonable to regard the sample of 10 men as representative of healthy adult males. Because the sample size is small ($n=10$), we must be willing to assume that the distribution of differences in MPF before and after diesel exposure is at least approximately normal. The following boxplot of the ten sample differences is fairly symmetric and there are no outliers, so it is not unreasonable to think that the population distribution could be at least approximately normal. These observations suggest that it is appropriate to use the paired-sample t test.

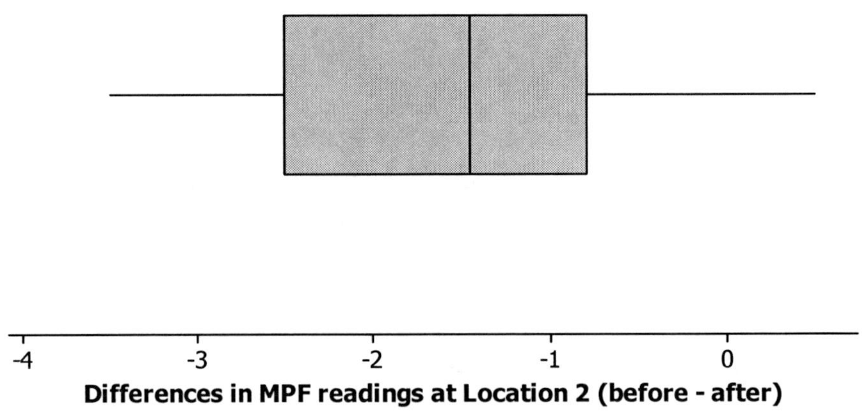

Calculations (C):

Summary statistics: $n = 10$, $\bar{x}_d = -1.540$ and $s_d = 1.186$

Test statistic: $t = \dfrac{-1.540 - 0}{\dfrac{1.186}{\sqrt{10}}} = -4.11$

Degrees of freedom when the null hypothesis is true is $\text{df} = n - 1 = 10 - 1 = 9$.

This is a lower-tailed test (the inequality in H_a is <), so the P-value is the total area under the t curve and to the left of –4.11 with 9 degrees of freedom. Therefore, the P-value is $P(t \le -4.11) = 0.001$.

Communicate Results (C):

Because the P-value of 0.001 is smaller than the selected significance level of $\alpha = 0.05$, we reject the null hypothesis. There is convincing evidence that the mean MPF at brain location 2 is higher after diesel exposure than before diesel exposure.

13.32: (a) Eosinophils. Virtually every child showed a reduction in eosinophils percentage, with many of the reductions by large amounts.

(b) FE_{NO}. While the majority seems to have shown a reduction, there are several who showed increases, and some of those by non-negligible amounts.

Section 13.2 Additional Exercises

13.37: Using the five-step process (HMC^3):

Hypotheses (H):

We want to determine if there is convincing evidence that the mean blood pressure is higher in a dental setting than in a medical setting. The population characteristics of interest are μ_{DS}, the mean blood pressure in a dental setting, μ_{MS}, the mean blood pressure in a medical setting, and $\mu_d = \mu_{DS} - \mu_{MS}$, the difference in the paired means. Test the hypotheses $H_0: \mu_d = 0$ versus $H_a: \mu_d > 0$.

Method (M):

Because the answers to the four key questions are (Q) hypothesis testing, (S) sample data, (T) one numerical variable, and (N) two paired samples,

we will consider the paired-samples t test as a potential method. The test statistic is $t=\frac{\bar{x}_d-\mu_0}{\frac{s_d}{\sqrt{n}}}$, and we will use a significance level of $\alpha=0.01$.

Check (C):

We are not given information about how the subjects were selected, so we will continue assuming that the subjects were either randomly selected or are representative of all such subjects. Because the sample size is large ($n=60$), we have satisfied the large sample requirement. These observations suggest that it is appropriate to use the paired-sample t test.

Calculations (C):

Summary statistics: $n=60$, $\bar{x}_d=4.47$ and $s_d=8.77$

Test statistic: $t=\frac{4.47-0}{\frac{8.77}{\sqrt{60}}}=3.95$

Degrees of freedom when the null hypothesis is true is $\text{df}=n-1=60-1=59$.

This is an upper-tailed test (the inequality in H_a is >), so the P-value is the total area under the t curve and to the right of 3.95 with 59 degrees of freedom. Therefore, the P-value is $P(t\geq 3.95)\approx 0$.

Communicate Results (C):

Because the P-value of approximately 0 is smaller than the selected significance level of $\alpha=0.01$, we reject the null hypothesis. There is convincing evidence that the mean blood pressure is higher in a dental setting than in a medical setting.

13.39: Using the five-step process (HMC^3):

Hypotheses (H):

We want to determine if there is convincing evidence that the mean memory score is higher after completion of the chess program. The population characteristics of interest are μ_{PRE}, the mean pre-program score on the memory test, μ_{POST}, the mean post-program score on the memory test, and $\mu_d = \mu_{PRE} - \mu_{POST}$, the difference in the paired means. Test the hypotheses $H_0 : \mu_d = 0$ versus $H_a : \mu_d < 0$.

Method (M):

Because the answers to the four key questions are (Q) hypothesis testing, (S) sample data, (T) one numerical variable, and (N) two paired samples, we will consider the paired-samples *t* test as a potential method. The test statistic is $t = \frac{\bar{x}_d - \mu_0}{\frac{s_d}{\sqrt{n}}}$, and we will use a significance level of $\alpha = 0.05$.

Check (C):

We are not given information about how the subjects were selected, so we will continue assuming that the subjects were either randomly selected or are representative of all such subjects. Because the sample size is small ($n = 12$), we must be willing to assume that the distribution of differences in memory test scores before and after the chess program is at least approximately normal. The following boxplot of the twelve sample differences is fairly symmetric and there are no outliers, so it is not unreasonable to think that the population distribution could be at least approximately normal. These observations suggest that it is appropriate to use the paired-sample *t* test.

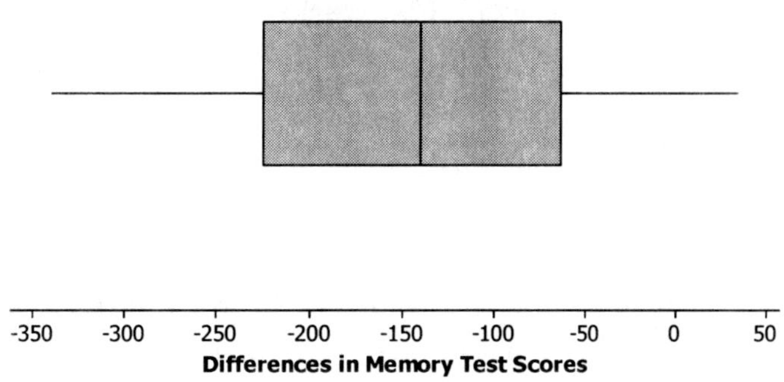

Calculations (C):

Summary statistics: $n = 12$, $\bar{x}_d = -144.6$ and $s_d = 109.7$

Test statistic: $t = \dfrac{-144.6 - 0}{\dfrac{109.7}{\sqrt{12}}} = -4.57$

Degrees of freedom when the null hypothesis is true is $\text{df} = n - 1 = 12 - 1 = 11$.

This is a lower-tailed test (the inequality in H_a is <), so the P-value is the total area under the t curve and to the left of –4.57 with 11 degrees of freedom. Therefore, the P-value is $P(t \leq -4.57) \approx 0$.

Communicate Results (C):

Because the P-value of approximately 0 is smaller than the selected significance level of $\alpha = 0.05$, we reject the null hypothesis. There is convincing evidence that the mean memory score is higher after completion of the chess program than before the program.

13.41: (a) Using the five-step process (HMC^3):

Hypotheses (H):

We want to determine if there is convincing evidence that the mean percentage of exams earning college credit at central coast high schools in 1997 and in 2002 were different. The population characteristics of interest are μ_{1997}, the mean percentage of exams earning college credit at central coast high schools in 1997, μ_{2002}, the mean percentage of exams earning college credit at central coast high schools in 2002, and $\mu_d = \mu_{1997} - \mu_{2002}$, the difference in the paired means. Test the hypotheses $H_0 : \mu_d = 0$ versus $H_a : \mu_d \neq 0$.

Method (M):

Because the answers to the four key questions are (Q) hypothesis testing, (S) sample data, (T) one numerical variable, and (N) two paired samples,

we will consider the paired-samples t test as a potential method. The test statistic is $t = \dfrac{\bar{x}_d - \mu_0}{\frac{s_d}{\sqrt{n}}}$, and we will use a significance level of $\alpha = 0.05$.

Check (C):

We are told that it is reasonable to regard these seven schools as a random sample of high schools located on the central coast of California. Because the sample size is small ($n = 7$), we must be willing to assume that the distribution of differences in the percentages of exams earning college credit in 1997 and 2002 is at least approximately normal. The following boxplot of the seven sample differences is fairly symmetric and there are no outliers, so it is not unreasonable to think that the population distribution could be at least approximately normal. These observations suggest that it is appropriate to use the paired-sample t test.

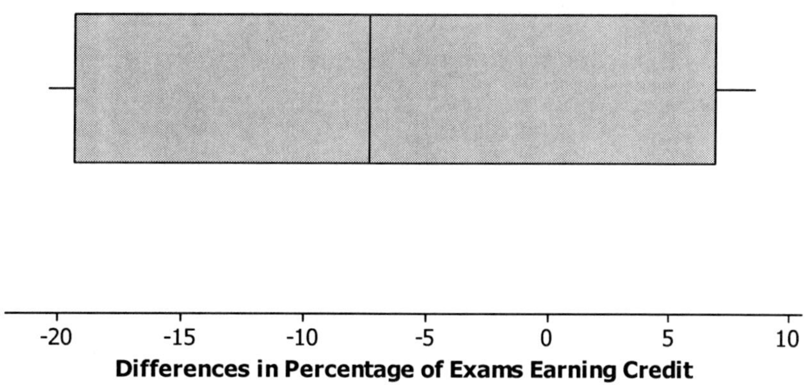

Calculations (C):

Summary statistics: $n = 7$, $\bar{x}_d = -5.24$ and $s_d = 12.05$

Test statistic: $t = \frac{-5.24 - 0}{\frac{12.05}{\sqrt{7}}} = -1.15$

Degrees of freedom when the null hypothesis is true is
$\text{df} = n - 1 = 7 - 1 = 6$.

This is a two-tailed test (the inequality in H_a is $\neq$), so the P-value is the total area under the t curve and to the left of -1.15 and to the right of 1.15 with 6 degrees of freedom. Therefore, using the symmetry of the t curve, the P-value is $2 \cdot P(t \leq -1.15) = 0.294$.

Communicate Results (C):

Because the P-value of 0.294 is larger than the selected significance level of $\alpha = 0.05$, we fail to reject the null hypothesis. There is not convincing evidence that the mean percentage of exams earning college credit at central coast high schools in 1997 and in 2002 were different.

(b) No. The sample consisted of schools from the central coast area only, and so the conclusion cannot be generalized to all California high schools.

(c) The boxplot of differences in percentage of students taking AP exams (below) shows that the distribution of differences contains an extreme outlier, and therefore we are not justified in using the paired-sample t test on this data set.

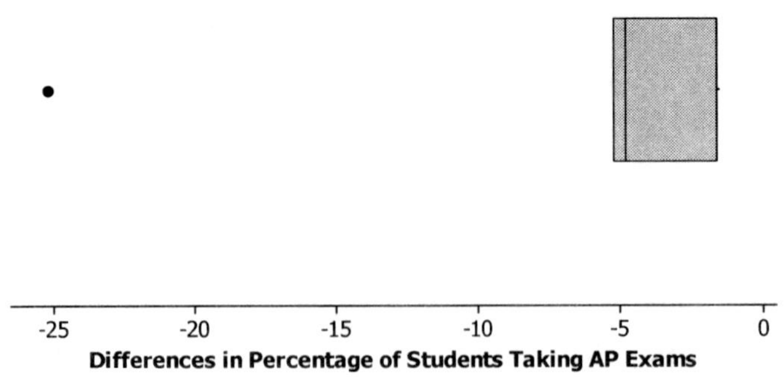

13.43: Using the five-step process (HMC^3):

Hypotheses (H):

We want to determine if there is convincing evidence that the mean cost-to-charge ratio for Oregon hospitals is lower for outpatient care than for inpatient care. The population characteristics of interest are μ_{IN}, the mean cost-to-charge ratio for inpatient care, μ_{OUT}, the mean cost-to-charge ratio for outpatient care, and $\mu_d = \mu_{IN} - \mu_{OUT}$, the difference in the paired means. Test the hypotheses $H_0 : \mu_d = 0$ versus $H_a : \mu_d > 0$.

Method (M):

Because the answers to the four key questions are (Q) hypothesis testing, (S) sample data, (T) one numerical variable, and (N) two paired samples, we will consider the paired-samples *t* test as a potential method. The test statistic is $t = \dfrac{\bar{x}_d - \mu_0}{\frac{s_d}{\sqrt{n}}}$, and we will use a significance level of $\alpha = 0.05$.

Check (C):

We are told that the study hospitals were randomly selected from all Oregon hospitals. Because the sample size is small ($n = 6$), we must be willing to assume that the distribution of differences in cost-to-charge ratios is at least approximately normal. The following boxplot of the six sample differences is fairly symmetric and there are no outliers, so it is not unreasonable to think that the population distribution could be at least approximately normal. These observations suggest that it is appropriate to use the paired-sample *t* test.

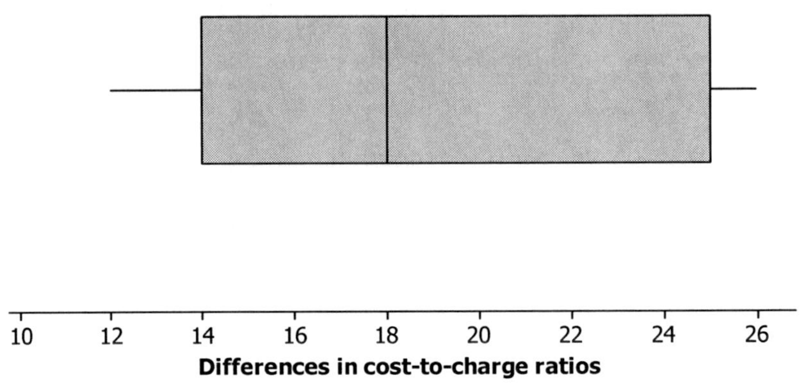

Calculations (C):

Summary statistics: $n = 6$, $\bar{x}_d = 18.83$ and $s_d = 5.67$

Test statistic: $t = \dfrac{18.83 - 0}{\dfrac{5.67}{\sqrt{6}}} = 8.13$

Degrees of freedom when the null hypothesis is true is $\text{df} = n - 1 = 6 - 1 = 5$.

This is an upper-tailed test (the inequality in H_a is >), so the P-value is the total area under the t curve and to the right of 8.13 with 5 degrees of freedom. Therefore, the P-value is $P(t \geq 8.13) \approx 0$.

Communicate Results (C):

Because the P-value of approximately 0 is smaller than the selected significance level of $\alpha = 0.05$, we reject the null hypothesis. There is convincing evidence that the mean cost-to-charge ratio for Oregon hospitals is lower for outpatient care than for inpatient care.

13.45: Using the five-step process (HMC^3):

Hypotheses (H):

We want to determine if there is convincing evidence that the mean number of seeds detected differs for the direct and stratified methods. The population characteristics of interest are μ_{DIRECT}, the mean number of seeds detected for the direct method, μ_{STRAT}, the mean number of seeds detected for the stratified method, and $\mu_d = \mu_{DIRECT} - \mu_{STRAT}$, the difference in the paired means. Test the hypotheses $H_0 : \mu_d = 0$ versus $H_a : \mu_d \neq 0$.

Method (M):

Because the answers to the four key questions are (Q) hypothesis testing, (S) sample data, (T) one numerical variable, and (N) two paired samples, we will consider the paired-samples *t* test as a potential method. The test statistic is $t = \dfrac{\bar{x}_d - \mu_0}{\dfrac{s_d}{\sqrt{n}}}$, and we will use a significance level of $\alpha = 0.05$.

Check (C):

We do not know if the soil specimens were randomly selected or if they are representative of all possible specimens, but we will continue as if this condition has been satisfied. Because the sample size is less than 30 ($n = 27$), we must be willing to assume that the distribution of differences in methods is at least approximately normal. The following boxplot of the 27 sample differences is skewed left and there are four outliers, so it is not reasonable to think that the population distribution could be at least approximately normal. However, we will proceed with caution.

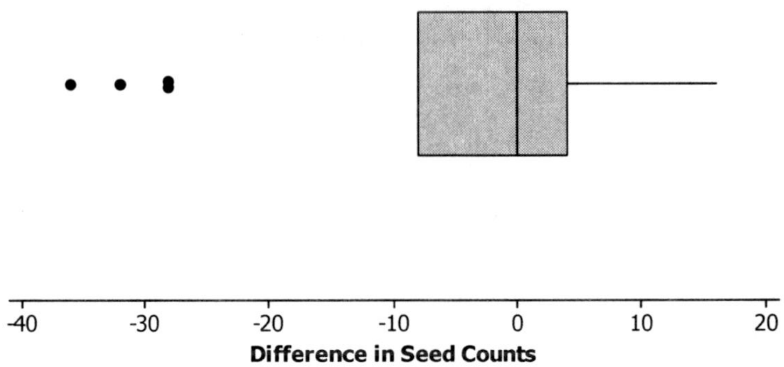

Calculations (C):

Summary statistics: $n = 27$, $\bar{x}_d = -3.41$ and $s_d = 13.25$

Test statistic: $t = \dfrac{-3.41 - 0}{\dfrac{13.25}{\sqrt{27}}} = -1.34$

Degrees of freedom when the null hypothesis is true is $\text{df} = n - 1 = 27 - 1 = 26$.

This is a two-tailed test (the inequality in H_a is $\neq$), so the P-value is the total area under the t curve and to the left of -1.34 and to the right of 1.34 with 26 degrees of freedom. Therefore, using the symmetry of the t curve, the P-value is $2 \cdot P(t \le -1.34) = 0.192$.

Communicate Results (C):

Because the P-value of 0.192 is greater than the selected significance level of $\alpha = 0.05$, we fail to reject the null hypothesis. There is not convincing evidence that the mean number of seeds detected differs for the two methods.

Section 13.3 Exercise Set 1

13.46: Using the five-step process (EMC^3):

Estimate (E):

We want to estimate the difference between the mean number of hours per day spent using electronic media in 2009 and 1999. The population characteristics of interest are μ_{2009}, the mean number of hours per day spent using electronic media in 2009, μ_{1999}, the mean number of hours per day spent using electronic media in 1999, and $\mu_{2009} - \mu_{1999}$, the difference in population means.

Method (M):

Because the answers to the four key questions are (Q) estimation, (S) sample data, (T) one numerical variable, and (N) two independent samples, we will consider constructing a two-sample *t* confidence interval using a confidence level of 98%.

Check (C):

We are told that the samples were selected to be representative of the two populations, and, given that the samples were selected 10 years apart, we will assume that the samples are independent. In addition, because both of the sample sizes are smaller than 30 ($n_{2009} = 15$ and $n_{1999} = 15$), we need to be willing to assume that the distributions are approximately normal for each of the two populations. The boxplots shown below constructed using the sample data are roughly symmetric and since there are no outliers in either sample, the assumption of normality is justified. Therefore, the necessary conditions have been satisfied and it is reasonable to construct a two-sample *t* confidence interval.

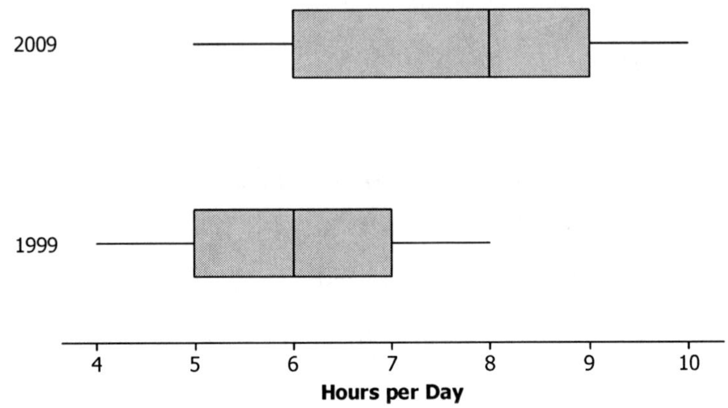

Calculations (C):

Summary statistics: $n_{2009} = 15$, $\bar{x}_{2009} = 7.6$ hours, $s_{2009} = 1.595$ hours, $n_{1999} = 15$, $\bar{x}_{1999} = 5.933$ hours, and $s_{1999} = 1.100$ hours

Degrees of freedom: $V_{2009} = \dfrac{s_{2009}^2}{n_{2009}} = \dfrac{(1.595)^2}{15} = 0.169602$,

$V_{1999} = \dfrac{s_{1999}^2}{n_{1999}} = \dfrac{(1.100)^2}{15} = 0.080667$ and

$$\text{df} = \frac{(V_{2009} + V_{1999})^2}{\dfrac{V_{2009}^2}{n_{2009}-1} + \dfrac{V_{1999}^2}{n_{1999}-1}} = \frac{(0.169602 + 0.080667)^2}{\dfrac{(0.169602)^2}{15-1} + \dfrac{(0.080667)^2}{15-1}} = 24.8606.$$

The t critical value found with technology for 98% confidence and using 24.8606 degrees of freedom is 2.486.

$$(\overline{x}_{2009} - \overline{x}_{1999}) \pm (t \text{ critical value})\sqrt{\frac{s_{2009}^2}{n_{2009}} + \frac{s_{1999}^2}{n_{1999}}}$$

$$(7.6 - 5.933) \pm (2.486)\sqrt{\frac{1.595^2}{15} + \frac{1.100^2}{15}}$$

$$1.667 \pm (2.486)0.5003$$

$$1.667 \pm 1.244$$

$$(0.423, 2.911)$$

Communicate Results (C):

Interpret Confidence Interval: We can be 98% confident that the difference between the mean number of hours spent per day using electronic media in 2009 and 1999 is between 0.423 and 2.911 hours. Both endpoints of this interval are positive, so you can estimate that the mean number of hours spent per day using electronic media in 2009 is greater than the mean number of hours spent per day using electronic media in 1999 by somewhere between 0.423 and 2.910 hours.

Interpret Confidence Level: The method used to construct this interval estimate is successful in capturing the actual difference in population means about 98% of the time.

13.47: Using the five-step process (EMC^3):

Estimate (E):

We want to estimate the difference between the mean daily calorie intake for teens who do eat fast food on a typical day and those who do not. The population characteristics of interest are μ_{FAST}, the mean daily calorie intake for teens who eat fast food, μ_{NOT}, the mean daily calorie intake for teens who do not eat fast food, and $\mu_{NOT} - \mu_{FAST}$, the difference in population means.

Method (M):

Because the answers to the four key questions are (Q) estimation, (S) sample data, (T) one numerical variable, and (N) two independent samples, we will consider constructing a two-sample *t* confidence interval using a confidence level of 95%.

Check (C):

We are told that the samples were selected to be representative of the two populations. In addition, because both of the sample sizes are large ($n_{FAST} = 413$ and $n_{NOT} = 663$), we do not need to worry about whether the distributions are normal for each of the two populations. Therefore, the necessary conditions have been satisfied and it is reasonable to construct a two-sample *t* confidence interval.

Calculations (C):

Summary statistics: $n_{FAST} = 413$, $\bar{x}_{FAST} = 2,637$ calories, $s_{FAST} = 1,138$ calories, $n_{NOT} = 663$, $\bar{x}_{NOT} = 2,258$ calories, and $s_{NOT} = 1,519$ calories

Degrees of freedom: $V_{FAST} = \frac{s_{FAST}^2}{n_{FAST}} = \frac{(1,138)^2}{413} = 3,135.7$,

$V_{NOT} = \frac{s_{NOT}^2}{n_{NOT}} = \frac{(1,519)^2}{663} = 3,480.18$ and

$$\text{df} = \frac{(V_{FAST} + V_{NOT})^2}{\frac{V_{FAST}^2}{n_{FAST} - 1} + \frac{V_{NOT}^2}{n_{NOT} - 1}} = \frac{(3,135.7 + 3,480.18)^2}{\frac{(3,135.7)^2}{413-1} + \frac{(3,480.18)^2}{663-1}} = 1038.16.$$

The *t* critical value found with technology for 95% confidence and using 1038.16 degrees of freedom is 1.96225.

$$(\bar{x}_{NOT} - \bar{x}_{FAST}) \pm (t \text{ critical value})\sqrt{\frac{s^2_{FAST}}{n_{FAST}} + \frac{s^2_{NOT}}{n_{NOT}}}$$

$$(2,258 - 2,637) \pm (1.96225)\sqrt{\frac{1,138^2}{413} + \frac{1,519^2}{663}}$$

$$-379 \pm (1.96225)81.3381$$

$$-379 \pm 159.606$$

$$(-538.606, -219.394)$$

Communicate Results (C):

Interpret Confidence Interval: We can be 95% confident that the mean daily calorie intake for teens who do not eat fast food on a typical day minus the mean daily calorie intake for teens who do eat fast food on a typical day is between –538.606 and –219.394 calories. Both endpoints of the confidence interval are negative, so you can estimate that the mean calorie intake for teens who eat fast food is greater than the mean calorie intake for teens who do not eat fast food by somewhere between 219.394 and 538.606 calories.

Interpret Confidence Level: The method used to construct this interval estimate is successful in capturing the actual difference in population means about 95% of the time.

13.48: Using the five-step process (EMC^3):

Estimate (E):

We want to estimate the difference between the mean MPF at brain location 1 before and after exposure to diesel exhaust. The population characteristics of interest are μ_{1B}, the mean MPF at brain location 1 before diesel exposure, μ_{1A}, the mean MPF at brain location 1 after diesel exposure, and $\mu_d = \mu_{1B} - \mu_{1A}$, the difference in the paired means.

Method (M):

Because the answers to the four key questions are (Q) estimation, (S) sample data, (T) one numerical variable, and (N) two paired samples, we will consider constructing a paired-samples *t* confidence interval using a confidence level of 90%.

Check (C):

We are told to assume that it is reasonable to regard the sample of 10 men as representative of healthy adult males. Because the sample size is small ($n = 10$), we must be willing to assume that the distribution of differences in MPF before and after diesel exposure is at least approximately normal. The following boxplot of the ten sample differences is fairly symmetric and there are no outliers, so it is not unreasonable to think that the population distribution could be at least approximately normal. These observations suggest that it is appropriate to construct a paired-sample *t* interval.

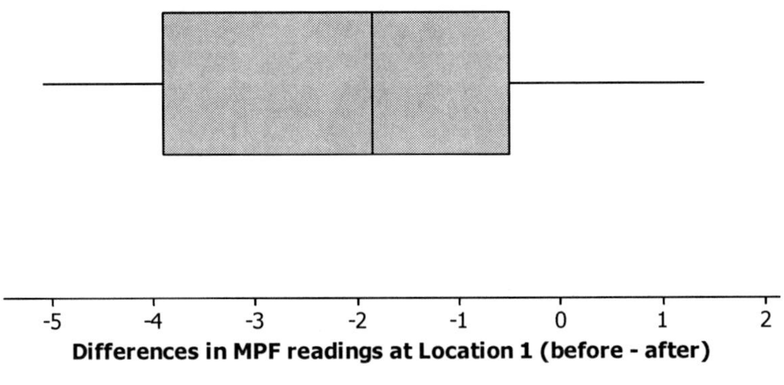

Calculations (C):

Summary statistics: $n = 10$, $\bar{x}_d = -1.930$ and $s_d = 1.965$

Degrees of freedom: df $= n - 1 = 10 - 1 = 9$

The *t* critical value found with technology for 90% confidence and using 9 degrees of freedom is 1.83311.

$$\bar{x}_d \pm (t \text{ critical value})\frac{s_d}{\sqrt{n}}$$

$$-1.930 \pm (1.83311)\frac{1.965}{\sqrt{10}}$$

$$-1.930 \pm (1.83311)0.621388$$

$$-1.930 \pm 1.13907$$

$$(-3.069, -0.791)$$

Communicate Results (C):

Interpret Confidence Interval: We can be 90% confident that the difference between the mean MPF at brain location 1 before and after exposure to diesel exhaust is between –3.069 and –0.791 Hz. Both endpoints of this interval are negative, so you can estimate that the mean MPF at brain location 1 after exposure to diesel exhaust is greater than the mean MPF at brain location 1 before exposure to diesel exhaust by somewhere between 0.791 and 3.069 Hz.

Interpret Confidence Level: The method used to construct this interval estimate is successful in capturing the actual difference in population means about 90% of the time.

13.49: Using the five-step process (EMC^3):

Estimate (E):

We want to estimate the difference between the mean MPF at brain location 2 before and after exposure to diesel exhaust. The population characteristics of interest are μ_{2B}, the mean MPF at brain location 2 before diesel exposure, μ_{2A}, the mean MPF at brain location 2 after diesel exposure, and $\mu_d = \mu_{2B} - \mu_{2A}$, the difference in the paired means.

Method (M):

Because the answers to the four key questions are (Q) estimation, (S) sample data, (T) one numerical variable, and (N) two paired samples, we will consider constructing a paired-samples t confidence interval using a confidence level of 90%.

Check (C):

We are told to assume that it is reasonable to regard the sample of 10 men as representative of healthy adult males. Because the sample size is small ($n = 10$), we must be willing to assume that the distribution of differences in MPF before and after diesel exposure is at least approximately normal. The following boxplot of the ten sample differences is fairly symmetric and there are no outliers, so it is not unreasonable to think that the population distribution could be at least approximately normal. These observations suggest that it is appropriate to construct a paired-sample t interval.

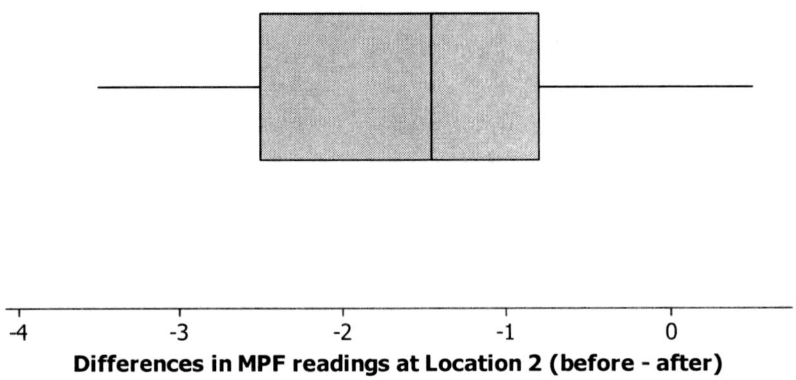

Calculations (C):

Summary statistics: $n = 10$, $\bar{x}_d = -1.540$ and $s_d = 1.186$

Degrees of freedom: df $= n - 1 = 10 - 1 = 9$

The t critical value found with technology for 90% confidence and using 9 degrees of freedom is 1.83311.

$$\bar{x}_d \pm (t \text{ critical value})\frac{s_d}{\sqrt{n}}$$

$$-1.540 \pm (1.83311)\frac{1.186}{\sqrt{10}}$$

$$-1.540 \pm (1.83311)0.375046$$

$$-1.540 \pm 0.687501$$

$$(-2.228, -0.852)$$

Communicate Results (C):

Interpret Confidence Interval: We can be 90% confident that the difference between the mean MPF at brain location 2 before and after exposure to diesel exhaust is between –2.228 and –0.852 Hz. Both endpoints of this interval are negative, so you can estimate that the mean MPF at brain location 2 after exposure to diesel exhaust is greater than the mean MPF at brain location 2 before exposure to diesel exhaust by somewhere between 0.852 and 2.228 Hz.

Interpret Confidence Level: The method used to construct this interval estimate is successful in capturing the actual difference in population means about 90% of the time.

13.50: (a) There is a single sample of girls with two data values for each girl.

(b) Using the five-step process (EMC^3):

Estimate (E):

We want to estimate the mean difference between the number of science classes a girl intends to take and the number she thinks boys should take. The population characteristics of interest are μ_G, the mean number of science classes she intends to take, μ_B, the mean number of science classes she thinks boys should take, and $\mu_d = \mu_G - \mu_B$, the difference in the paired means.

Method (M):

Because the answers to the four key questions are (Q) estimation, (S) sample data, (T) one numerical variable, and (N) two paired samples, we will consider constructing a paired-samples t confidence interval using a confidence level of 95%.

Check (C):

We are told that a random sample of children in grades 4, 5, and 6 was taken. Because the sample size is large ($n=224$), the large sample condition is met. These observations suggest that it is appropriate to construct a paired-sample t interval.

Calculations (C):

Summary statistics: $n=224$, $\bar{x}_d=-0.83$ and $s_d=1.51$

Degrees of freedom: df $=n-1=224-1=223$

The t critical value found with technology for 95% confidence and using 224 degrees of freedom is 1.97066.

$$\bar{x}_d \pm (t \text{ critical value})\frac{s_d}{\sqrt{n}}$$
$$-0.83 \pm (1.97066)\frac{1.51}{\sqrt{224}}$$
$$-0.83 \pm (1.97066)0.100891$$
$$-0.83 \pm 0.198822$$
$$(-1.029, -0.631)$$

Communicate Results (C):

Interpret Confidence Interval: We can be 95% confident that the mean difference between the number of science courses a girl intends to take and the number of science courses she thinks a boy should take is between –1.029 and –0.631. Both endpoints of this interval are negative, so you can estimate that the mean number of science courses a girl intends to take is less than the number of science courses she thinks a boy should take by somewhere between 0.631 and 1.029.

Interpret Confidence Level: The method used to construct this interval estimate is successful in capturing the actual difference in population means about 95% of the time.

Section 13.3 Additional Exercises

13.57: Using the five-step process (EMC^3):

Estimate (E):

We want to estimate the difference in mean GPA for students at the University of Central Florida who are employed and students who are not employed. The population characteristics of interest are μ_E, the mean GPA of students who are employed, μ_{NE}, the mean GPA of students who are not employed, and $\mu_E - \mu_{NE}$, the difference in population means.

Method (M):

Because the answers to the four key questions are (Q) estimation, (S) sample data, (T) one numerical variable, and (N) two independent samples, we will consider constructing a two-sample *t* confidence interval using a confidence level of 95%.

Check (C):

We are told that the samples were selected at random from the populations of working and nonworking students and, as such, are independent of each other. In addition, because both of the sample sizes are greater than 30 ($n_E = 184$ and $n_{NE} = 114$), the large sample condition has been satisfied. Therefore, the necessary conditions have been satisfied and it is reasonable to construct a two-sample *t* confidence interval.

Calculations (C):

Summary statistics: $n_E = 184$, $\bar{x}_E = 3.12$, $s_E = 0.485$, $n_{NE} = 114$, $\bar{x}_{NE} = 3.23$, and $s_{NE} = 0.524$

Degrees of freedom: $V_E = \frac{s_E^2}{n_E} = \frac{(0.485)^2}{184} = 0.001278$,

$$V_{NE} = \frac{s_{NE}^2}{n_{NE}} = \frac{(0.524)^2}{114} = 0.002409 \text{ and}$$

$$\text{df} = \frac{(V_E + V_{NE})^2}{\frac{V_E^2}{n_E - 1} + \frac{V_{NE}^2}{n_{NE} - 1}} = \frac{(0.001278 + 0.002409)^2}{\frac{(0.001278)^2}{184-1} + \frac{(0.002409)^2}{114-1}} = 225.508.$$

The t critical value found with technology for 95% confidence and using 225.508 degrees of freedom is 1.97054.

$$(\bar{x}_E - \bar{x}_{NE}) \pm (t \text{ critical value})\sqrt{\frac{s_E^2}{n_E} + \frac{s_{NE}^2}{n_{NE}}}$$

$$(3.12 - 3.23) \pm (1.97054)\sqrt{\frac{0.485^2}{184} + \frac{0.524^2}{114}}$$

$$-0.11 \pm (1.97054)0.06072$$

$$-0.11 \pm 0.119652$$

$$(-0.230, 0.010)$$

Communicate Results (C):

Interpret Confidence Interval: We can be 95% confident that the difference in mean GPA for students at the University of Central Florida who are employed and students who are not employed is between –0.230 and 0.010. Because 0 is included in the interval, it is possible that there is no difference in the mean GPA for working and nonworking students.

Interpret Confidence Level: The method used to construct this interval estimate is successful in capturing the actual difference in population means about 95% of the time.

13.59: Using the five-step process (EMC^3):

Estimate (E):

We want to estimate the difference in mean number of words remembered after 1 hour and after 24 hours. The population characteristics of interest are μ_1, the mean number words remembered after 1 hour, μ_{24}, the mean number of words remembered after 24 hours, and $\mu_d = \mu_1 - \mu_{24}$, the difference in the paired means.

Method (M):

Because the answers to the four key questions are (Q) estimation, (S) sample data, (T) one numerical variable, and (N) two paired samples, we will consider constructing a paired-samples *t* confidence interval using a confidence level of 90%.

Check (C):

We are told that a random sample of students in a large psychology class was taken. Because the sample size is small ($n = 8$), we must be willing to assume that the distribution of differences in the number of words remembered after 1 hour and after 24 hours is approximately normal. The following boxplot of the eight sample differences is positively skewed and there is one outlier, so it is not reasonable to think that the population distribution could be at least approximately normal. However, we will proceed with caution.

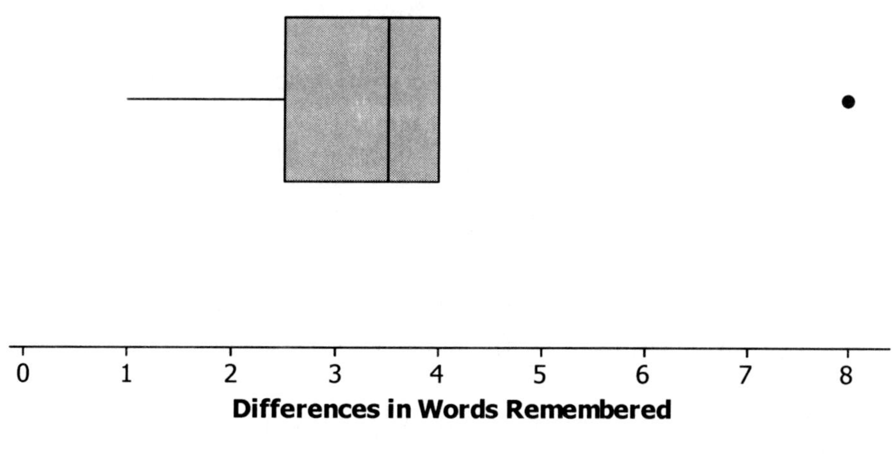

Calculations (C):

Summary statistics: $n = 8$, $\bar{x}_d = 3.625$ and $s_d = 2.066$

Degrees of freedom: $\text{df} = n-1 = 8-1 = 7$

The t critical value found with technology for 90% confidence and using 7 degrees of freedom is 1.89458.

$$\bar{x}_d \pm (t \text{ critical value})\frac{s_d}{\sqrt{n}}$$
$$3.625 \pm (1.89458)\frac{2.066}{\sqrt{8}}$$
$$3.625 \pm (1.89458)0.730441$$
$$3.625 \pm 1.38388$$
$$(2.241, 5.009)$$

Communicate Results (C):

Interpret Confidence Interval: We can be 90% confident that the difference in mean number of words remembered after 1 hour and after 24 hours is between 2.241 and 5.009. Both endpoints of this interval are positive, so you can estimate that the mean number words remembered after 1 hour is greater than the mean number of words remembered after 24 hours by somewhere between 2.241 and 5.009.

Interpret Confidence Level: The method used to construct this interval estimate is successful in capturing the actual difference in population means about 90% of the time.

Are You Ready to Move On? Chapter 13 Review Exercises

13.61: (a) $H_0: \mu_1 - \mu_2 = 10$ versus $H_a: \mu_1 - \mu_2 > 10$

(b) $H_0: \mu_1 - \mu_2 = -10$ versus $H_a: \mu_1 - \mu_2 < -10$

13.63: Scenario 1: Testing a difference in population means is not appropriate because the variable in this study is categorical.

Scenario 2: Testing a difference in population means is not appropriate because this study has one sample whose mean is being compared with a national average.

Scenario 3: Testing a difference in population means is appropriate because this study has two independent samples and the variable is numerical.

13.65: Using the five-step process (HMC^3):

Hypotheses (H):

We want to determine if there is convincing evidence that there is a difference in mean sodium content for meal purchases at Burger King and meal purchases at McDonald's. The population characteristics of interest are μ_B, the mean sodium content for meal purchases at Burger King, μ_M, the mean sodium content for meal purchases at McDonald's, and $\mu_B - \mu_M$, the difference in population means. Test the hypotheses $H_0: \mu_B - \mu_M = 0$ versus $H_a: \mu_B - \mu_M \neq 0$.

Method (M):

Because the answers to the four key questions are (Q) hypothesis testing, (S) sample data, (T) one numerical variable, and (N) two independent samples, we will consider a two-sample *t* test for a difference in population means. The test statistic is: $t = \dfrac{(\bar{x}_B - \bar{x}_M) - (\mu_B - \mu_M)}{\sqrt{\dfrac{s_B^2}{n_B} + \dfrac{s_M^2}{n_M}}}$, and we will use a significance level of $\alpha = 0.05$.

Check (C):

We are told that the samples were randomly selected and independent. In addition, because both of the sample sizes are larger than 30 ($n_B = 850$ and $n_M = 2{,}107$), the large sample condition is met. The three required conditions are satisfied.

Calculations (C):

Summary statistics: $n_B = 850$, $\bar{x}_B = 1{,}685$ mg, $s_B = 828$ mg, $n_M = 2{,}107$, $\bar{x}_M = 1{,}477$ mg, and $s_M = 812$ mg

Test statistic: $t = \dfrac{(1{,}685 - 1{,}477) - 0}{\sqrt{\dfrac{(828)^2}{850} + \dfrac{(812)^2}{2{,}107}}} = 6.22$

Degrees of freedom: $V_B = \dfrac{s_B^2}{n_B} = \dfrac{(828)^2}{850} = 806.569$,

$V_M = \dfrac{s_M^2}{n_M} = \dfrac{(812)^2}{2{,}107} = 312.93$ and

$\text{df} = \dfrac{(V_B + V_M)^2}{\dfrac{V_B^2}{n_B - 1} + \dfrac{V_M^2}{n_M - 1}} = \dfrac{(806.569 + 312.93)^2}{\dfrac{(806.569)^2}{850-1} + \dfrac{(312.93)^2}{2{,}107-1}} = 1{,}542.01$, truncated to

$\text{df} = 1{,}542$.

This is a two-tailed test (the inequality in H_a is $\neq$), so the *P*-value is the area under the *t* curve and to the left of –6.22 and to the right of 6.22 with 1,542 degrees of freedom. Therefore, using the symmetry of the *t* curve, the *P*-value is $2 \cdot P(t \le -6.22) \approx 0$.

Communicate Results (C):

Because the *P*-value of approximately 0 is less than the selected significance level of $\alpha = 0.05$, we reject the null hypothesis. There is convincing evidence that the mean sodium content is not the same for meal purchases at Burger King and McDonalds.

13.67: Using the five-step process (HMC^3):

Hypotheses (H):

We want to determine if there is convincing evidence that the mean energy expenditure using the conventional shovel exceeds that using the perforated shovel. The population characteristics of interest are μ_C, the mean energy expenditure using the conventional shovel, μ_P, the mean energy expenditure using the perforated shovel, and $\mu_d = \mu_C - \mu_P$, the difference in the paired means. Test the hypotheses $H_0 : \mu_d = 0$ versus $H_a : \mu_d > 0$.

Method (M):

Because the answers to the four key questions are (Q) hypothesis testing, (S) sample data, (T) one numerical variable, and (N) two paired samples, we will consider the paired-samples t test as a potential method. The test statistic is $t = \dfrac{\bar{x}_d - \mu_0}{\frac{s_d}{\sqrt{n}}}$, and we will use a significance level of $\alpha = 0.05$.

Check (C):

We are not told how the workers were selected, or whether they are representative of the population of all workers. We will continue with caution. Because the sample size is small ($n = 13$), we must be willing to assume that the distribution of differences in energy expenditures is at least approximately normal. The following boxplot of the thirteen sample differences is only somewhat skewed and there are no outliers, so it is not unreasonable to think that the population distribution could be at least approximately normal.

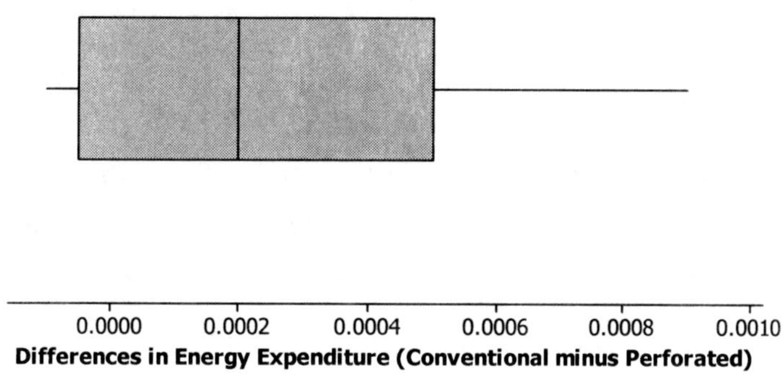

Calculations (C):

Summary statistics: $n = 13$, $\bar{x}_d = 0.000246$ and $s_d = 0.000331$

Test statistic: $t = \dfrac{0.000246 - 0}{\frac{0.000331}{\sqrt{13}}} = 2.68$

Degrees of freedom when the null hypothesis is true is
$\text{df} = n-1 = 13-1 = 12$.

This is an upper-tailed test (the inequality in H_a is >), so the P-value is the total area under the t curve and to the right of 2.68 with 12 degrees of freedom. Therefore, the P-value is $P(t \geq 2.68) = 0.010$.

Communicate Results (C):

Because the P-value of 0.010 is smaller than the selected significance level of $\alpha = 0.05$, we reject the null hypothesis. There is convincing evidence that the mean energy expenditure is greater for the conventional shovel than for the perforated shovel.

13.69: (a) Using the five-step process (HMC³):

Hypotheses (H):

We want to determine if there is convincing evidence that evidence that, on average, male online daters overstate their height in online dating profiles. The population characteristics of interest are μ_P, the mean online profile height for men, μ_A, the mean actual height for men, and $\mu_d = \mu_P - \mu_A$, the difference in the paired means. Test the hypotheses $H_0 : \mu_d = 0$ versus $H_a : \mu_d > 0$.

Method (M):

Because the answers to the four key questions are (Q) hypothesis testing, (S) sample data, (T) one numerical variable, and (N) two paired samples, we will consider the paired-samples t test as a potential method. The test statistic is $t = \dfrac{\bar{x}_d - \mu_0}{\frac{s_d}{\sqrt{n}}}$, and we will use a significance level of $\alpha = 0.05$.

Check (C):

We are told to assume that the sample is representative of male online daters. Because the sample size is greater than 30 ($n = 40$), the large sample condition has been satisfied. These observations suggest that it is appropriate to use the paired-sample t test.

Calculations (C):

Summary statistics: $n = 40$, $\bar{x}_d = 0.57$ inches, and $s_d = 0.81$ inches

Test statistic: $t = \dfrac{0.57 - 0}{\dfrac{0.81}{\sqrt{40}}} = 4.451$

Degrees of freedom when the null hypothesis is true is
$\text{df} = n - 1 = 40 - 1 = 39$.

This is an upper-tailed test (the inequality in H_a is >), so the P-value is the total area under the t curve and to the right of 4.451 with 39 degrees of freedom. Therefore, the P-value is $P(t \geq 4.451) \approx 0$.

Communicate Results (C):

Because the P-value of approximately 0 is smaller than the selected significance level of $\alpha = 0.05$, we reject the null hypothesis. There is convincing evidence that, on average, male online daters overstate their height.

(b) Using the five-step process (EMC^3):

Estimate (E):

We want to estimate the difference between the mean online profile height and mean actual height for female online daters. The population characteristics of interest are μ_P, the mean online profile height for female online daters, μ_A, the mean actual height for female online daters, and $\mu_d = \mu_P - \mu_A$, the difference in the paired means.

Method (M):

Because the answers to the four key questions are (Q) estimation, (S) sample data, (T) one numerical variable, and (N) two paired samples, we will consider constructing a paired-samples t confidence interval using a confidence level of 95%.

Check (C):

We are told to assume that the sample is representative of female online daters. Because the sample size is large ($n = 40$), the large sample condition is met. These observations suggest that it is appropriate to construct a paired-sample t interval.

Calculations (C):

Summary statistics: $n = 40$, $\bar{x}_d = 0.03$ inches and $s_d = 0.75$ inches

Degrees of freedom: df $= n - 1 = 40 - 1 = 39$

The t critical value found with technology for 95% confidence and using 39 degrees of freedom is 2.02269.

$$\bar{x}_d \pm (t \text{ critical value}) \frac{s_d}{\sqrt{n}}$$

$$0.03 \pm (2.02269) \frac{0.75}{\sqrt{40}}$$

$$0.03 \pm (2.02269) 0.118585$$

$$0.03 \pm 0.239862$$

$$(-0.210, 0.270)$$

Communicate Results (C):

Interpret Confidence Interval: We can be 95% confident that the mean difference between the online profile reported height and actual height is between –0.210 and 0.270 inches. Because 0 is included in the interval, it is possible that there is no difference in the mean online profile reported height and the mean actual height for women.

Interpret Confidence Level: The method used to construct this interval estimate is successful in capturing the actual difference in population means about 95% of the time.

(c) Using the five-step process (HMC^3):

Hypotheses (H):

We want to determine if there is convincing evidence that the mean height difference (profile – actual) for male online daters is greater than the mean height difference (profile – actual) for female online daters. The population characteristics of interest are μ_m, the mean height difference (profile – actual) for male online daters, μ_f, the mean height difference (profile – actual) for female online daters, and $\mu_m - \mu_f$, the difference in population means. Test the hypotheses $H_0 : \mu_m - \mu_f = 0$ versus $H_a : \mu_m - \mu_f > 0$.

Method (M):

Because the answers to the four key questions are (Q) hypothesis testing, (S) sample data, (T) one numerical variable, and (N) two independent samples, we will consider a two-sample t test for a difference in population means. The test statistic is: $t = \dfrac{(\bar{x}_m - \bar{x}_f) - (\mu_m - \mu_f)}{\sqrt{\dfrac{s_m^2}{n_m} + \dfrac{s_f^2}{n_f}}}$, and we will use a significance level of $\alpha = 0.05$.

Check (C):

We are told to assume that the samples are representative of the populations of male and female online daters. In addition, because both of the sample sizes are larger than 30 ($n_m = 40$ and $n_f = 40$), the large sample condition is met. The required conditions are satisfied.

Calculations (C):

Summary statistics: $n_m = 40$, $\bar{x}_m = 0.57$ inches, $s_m = 0.81$ inches, $n_f = 40$, $\bar{x}_f = 0.03$ inches, and $s_f = 0.75$ inches

Test statistic: $t = \dfrac{(0.57 - 0.03) - 0}{\sqrt{\dfrac{(0.81)^2}{40} + \dfrac{(0.75)^2}{40}}} = 3.094$

Degrees of freedom: $V_m = \frac{s_m^2}{n_m} = \frac{(0.81)^2}{40} = 0.016403$,

$V_f = \frac{s_f^2}{n_f} = \frac{(0.75)^2}{40} = 0.014063$ and

$$\text{df} = \frac{(V_m + V_f)^2}{\frac{V_m^2}{n_m - 1} + \frac{V_f^2}{n_f - 1}} = \frac{(0.016403 + 0.014063)^2}{\frac{(0.016403)^2}{40-1} + \frac{(0.014063)^2}{40-1}} = 77.5426\text{, truncated to}$$

$\text{df} = 77$.

This is an upper-tailed test (the inequality in H_a is >), so the *P*-value is the area under the *t* curve and to the right of 3.094 with 77 degrees of freedom. Therefore, the *P*-value is $P(t \geq 3.094) = 0.001$.

Communicate Results (C):

Because the *P*-value of 0.001 is less than the selected significance level of $\alpha = 0.05$, we reject the null hypothesis. There is convincing evidence that the mean height difference is greater for males than for females.

(d) In Part (a), the male profile heights and the male actual heights are paired (according to which individual has the actual height and the height stated in the profile), and with paired samples, we use the paired-samples *t* test. In Part (c), we were dealing with two independent samples (the sample of males and the sample of females), and therefore, the two-sample *t* test was appropriate.

13.71: (a) Using the five-step process (EMC^3):

Estimate (E):

We want to estimate the mean difference between surface and subsoil pH for agricultural land in this county. The population characteristics of interest are μ_{SURF}, the mean surface pH for agricultural land in this county, μ_{SUB}, the mean subsurface pH for agricultural land in this county, and $\mu_d = \mu_{SURF} - \mu_{SUB}$, the difference in the paired means.

Method (M):

Because the answers to the four key questions are (Q) estimation, (S) sample data, (T) one numerical variable, and (N) two paired samples, we will consider constructing a paired-samples *t* confidence interval using a confidence level of 90%.

Check (C):

We are told that a random sample of locations was taken. Because the sample size is small ($n = 8$),we must be willing to assume that the distribution of differences in surface and subsoil pH levels is at least approximately normal. The following boxplot of the eight sample differences is somewhat skewed but there are no outliers, so it is not unreasonable to think that the population distribution could be at least approximately normal. These observations suggest that it is appropriate to use the paired-sample *t* test.

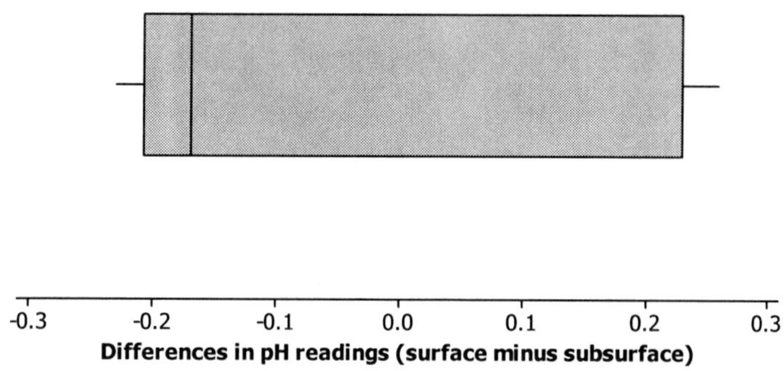

Calculations (C):

Summary statistics: $n = 8$, $\bar{x}_d = -0.0375$ and $s_d = 0.2213$

Degrees of freedom: df $= n - 1 = 8 - 1 = 7$

The *t* critical value found with technology for 90% confidence and using 7 degrees of freedom is 1.89458.

$$\bar{x}_d \pm (t \text{ critical value})\frac{s_d}{\sqrt{n}}$$

$$-0.0375 \pm (1.89458)\frac{0.2213}{\sqrt{8}}$$

$$-0.0375 \pm (1.89458)0.078241$$

$$-0.0375 \pm 0.148234$$

$$(-0.186, 0.111)$$

Communicate Results (C):

Interpret Confidence Interval: We can be 90% confident that the mean difference in pH for surface and subsoil is between –0.186 and 0.111. Because 0 is included in the interval, it is possible that there is no difference in mean pH.

Interpret Confidence Level: The method used to construct this interval estimate is successful in capturing the actual difference in population means about 90% of the time.

(b) You must assume that the distribution of differences across all locations is normal.

Chapter 14 Learning from Experiment Data

Section 14.1 Exercise Set 1

14.1: The mean quiz score for those in the texting group was enough lower than the mean for the no-texting group that a difference this large is not likely to have occurred by chance (due just to the random assignment) when there is no real difference between the treatment means.

14.2: The difference in mean calorie intake for the 4-hour and 8-hour sleep groups could have occurred by chance just due to the random assignment.

Section 14.1 Additional Exercises

14.5: (a) The two treatments are small fork and large fork.

(b) The mean amount of food for those in the small fork group was enough larger than the mean for the small fork group that a difference this large is not likely to have occurred by chance (due just to the random assignment) when there is no real difference between the treatment means.

14.7: The mean value assigned to the mug for those in the group that was given the mug was enough higher than the mean for the other group that a difference this large is not likely to have occurred by chance (due just to the random assignment) when there is no real difference between the treatment means.

Section 14.2 Exercise Set 1

14.9: Using the five-step process (HMC^3):

Hypotheses (H):

We want to determine if there is convincing evidence that the mean test score is higher for the new teaching method than for the traditional lecture method. The treatment means are μ_N, the mean test score for the new teaching method, μ_T, the mean test score for the traditional lecture method, and $\mu_N - \mu_T$, the difference in treatment means. Test the hypotheses $H_0 : \mu_N - \mu_T = 0$ versus $H_a : \mu_N - \mu_T > 0$.

Method (M):

Because the answers to the four key questions are (Q) hypothesis testing, (S) experiment data, (T) one numerical variable, and (N) two treatments, we will consider a two-sample *t* test for a difference in treatment means. The test statistic is $t = \frac{(\bar{x}_N - \bar{x}_T) - (\mu_N - \mu_T)}{\sqrt{\frac{s_N^2}{n_N} + \frac{s_T^2}{n_T}}}$, and we will use a significance level of $\alpha = 0.01$.

Check (C):

We are told that the students were randomly assigned to one of two experimental treatment groups. In addition, because both of the sample sizes are larger than 30 ($n_N = 425$ and $n_T = 425$), the large sample condition is met. The two required conditions are satisfied.

Calculations (C):

Summary statistics: $n_N = 425$, $\bar{x}_N = 74$, $s_N = 20$, $n_T = 425$, $\bar{x}_T = 41$, and $s_T = 24$

Test statistic: $t = \frac{(74-41)-0}{\sqrt{\frac{(20)^2}{425} + \frac{(24)^2}{425}}} = 21.78$

Degrees of freedom: $V_N = \frac{s_N^2}{n_N} = \frac{(20)^2}{425} = 0.941176$,

$V_T = \frac{s_T^2}{n_T} = \frac{(24)^2}{425} = 1.35529$ and

$\text{df} = \frac{(V_N + V_T)^2}{\frac{V_N^2}{n_N - 1} + \frac{V_T^2}{n_T - 1}} = \frac{(0.941176 + 1.35529)^2}{\frac{(0.941176)^2}{425-1} + \frac{(1.35529)^2}{425-1}} = 821.293$, truncated to $\text{df} = 821$.

This is an upper-tailed test (the inequality in H_a is >), so the P-value is the area under the t curve and to the right of 21.78 with 821 degrees of freedom. Therefore, the P-value is $P(t \geq 21.78) \approx 0$.

Communicate Results (C):

Because the P-value of approximately 0 is less than the selected significance level of $\alpha = 0.01$, we reject the null hypothesis. There is convincing evidence that the mean test score is higher for the new teaching method.

14.10: Using the five-step process (HMC^3):

Hypotheses (H):

We want to determine if there is convincing evidence to support the theory that learning the gesturing approach to solving problems of this type results in a higher mean number of correct responses. The treatment means are μ_G, the mean number of correct responses using the gesturing approach, μ_{NG}, the mean number of correct responses not using the gesturing approach, and $\mu_G - \mu_{NG}$, the difference in treatment means. Test the hypotheses $H_0 : \mu_G - \mu_{NG} = 0$ versus $H_a : \mu_G - \mu_{NG} > 0$.

Method (M):

Because the answers to the four key questions are (Q) hypothesis testing, (S) experiment data, (T) one numerical variable, and (N) two treatments, we will consider a two-sample t test for a difference in treatment means. The test statistic is $t = \dfrac{(\bar{x}_G - \bar{x}_{NG}) - (\mu_G - \mu_{NG})}{\sqrt{\dfrac{s_G^2}{n_G} + \dfrac{s_{NG}^2}{n_{NG}}}}$, and we will use a significance level of $\alpha = 0.05$.

Check (C):

We are told that the children were randomly assigned to one of two experimental treatment groups. In addition, because both of the sample sizes are larger than 30 ($n_G = 43$ and $n_{NG} = 42$), the large sample condition is met. The two required conditions are satisfied.

Calculations (C):

Summary statistics: $n_G = 43$, $\bar{x}_G = 2.2$, $s_G = 0.4$, $n_{NG} = 42$, $\bar{x}_{NG} = 1.3$, and $s_{NG} = 0.3$

Test statistic: $t = \dfrac{(2.2 - 1.3) - 0}{\sqrt{\dfrac{(0.4)^2}{43} + \dfrac{(0.3)^2}{42}}} = 11.753$

Degrees of freedom: $V_G = \dfrac{s_G^2}{n_G} = \dfrac{(0.4)^2}{43} = 0.003721$,

$V_{NG} = \dfrac{s_{NG}^2}{n_{NG}} = \dfrac{(0.3)^2}{42} = 0.002143$ and

$\text{df} = \dfrac{(V_G + V_{NG})^2}{\dfrac{V_G^2}{n_G - 1} + \dfrac{V_{NG}^2}{n_{NG} - 1}} = \dfrac{(0.003721 + 0.002143)^2}{\dfrac{(0.003721)^2}{43-1} + \dfrac{(0.002143)^2}{42-1}} = 77.855$, truncated to $\text{df} = 77$.

This is an upper-tailed test (the inequality in H_a is >), so the *P*-value is the area under the *t* curve and to the right of 11.753 with 77 degrees of freedom. Therefore, the *P*-value is $P(t \geq 11.753) \approx 0$.

Communicate Results (C):

Because the *P*-value of approximately 0 is less than the selected significance level of $\alpha = 0.01$, we reject the null hypothesis. There is convincing evidence that learning the gesturing approach to solving problems of this type results in a higher mean number of correct responses.

14.11: Using the five-step process (HMC^3):

Hypotheses (H):

We want to determine if there is convincing evidence to support the theory that the proportion of resumes receiving positive responses is higher for those resumes with "white-sounding" first names than for those resumes with "black-sounding" first names. The treatment proportions are p_W, the proportion of resumes with "white-sounding" first names receiving positive responses, p_B, the proportion of resumes with "black-sounding" first names receiving positive responses, and $p_W - p_B$, the difference in treatment proportions. Test the hypotheses $H_0: p_W - p_B = 0$ versus $H_a: p_W - p_B > 0$.

Method (M):

Because the answers to the four key questions are (Q) hypothesis testing, (S) experiment data, (T) one categorical variable, and (N) two treatments, we will consider a two-sample *z* test for a difference in treatment proportions. The test statistic is $z = \dfrac{\hat{p}_W - \hat{p}_B}{\sqrt{\dfrac{\hat{p}_c(1-\hat{p}_c)}{n_W} + \dfrac{\hat{p}_c(1-\hat{p}_c)}{n_B}}}$, and we will use a significance level of $\alpha = 0.05$.

Check (C):

We are told that the job ads were randomly assigned to either the "white-sounding" name group or the "black-sounding" name group. In addition, each of the treatment groups has at least 10 successes (250 resumes received responses in the "white-sounding" name group, and 167 resumes received responses in the "black-sounding" name group) and 10 failures (2,250 resumes did not receive responses in the "white-sounding" name group, and 2,333 resumes did not receive responses in the "black-sounding" name group). The two required conditions are satisfied.

Calculations (C):

Summary statistics: $\hat{p}_W = \dfrac{250}{2,500} = 0.10$ and $\hat{p}_B = \dfrac{167}{2,500} = 0.0668$

Combined estimate of the common proportion:

$$\hat{p}_c = \frac{n_W \hat{p}_W + n_B \hat{p}_B}{n_W + n_B} = \frac{(2,500)(0.10) + (2,500)(0.0668)}{2,500 + 2,500} = 0.0834$$

Test statistic: $$z = \frac{0.10 - 0.0668}{\sqrt{\dfrac{0.0834(1-0.0834)}{2,500} + \dfrac{0.0834(1-0.0834)}{2,500}}} = 4.245$$

This is an upper-tailed test (the inequality in H_a is >), so the P-value is the area under the z curve and to the right of 4.245. Therefore, the P-value is $P(z \geq 4.245) \approx 0$.

Communicate Results (C):

Because the P-value of approximately 0 is less than the selected significance level of $\alpha = 0.05$, we reject the null hypothesis. There is convincing evidence that the proportion receiving a response is higher for "white-sounding" first names than for "black-sounding" first names.

Section 14.2 Additional Exercises

14.15: Using the five-step process (HMC[3]):

Hypotheses (H):

We want to determine if there is convincing evidence to support the researchers claim that the mean quiz score for students who are permitted to text during the lecture is significantly lower than the mean quiz score for students who do not text. The treatment means are μ_T, the mean quiz score for students who are permitted to text during the lecture, μ_{NT}, the mean quiz score for students who are not permitted to text during the lecture, and $\mu_T - \mu_{NT}$, the difference in treatment means. Test the hypotheses $H_0: \mu_T - \mu_{NT} = 0$ versus $H_a: \mu_T - \mu_{NT} < 0$.

Method (M):

Because the answers to the four key questions are (Q) hypothesis testing, (S) experiment data, (T) one numerical variable, and (N) two treatments, we will consider a two-sample t test for a difference in treatment means. The test statistic is $t = \frac{(\bar{x}_T - \bar{x}_{NT}) - (\mu_T - \mu_{NT})}{\sqrt{\frac{s_T^2}{n_T} + \frac{s_{NT}^2}{n_{NT}}}}$, and we will use a significance level of $\alpha = 0.05$.

Check (C):

We are told that the undergraduate students were randomly assigned to one of two experimental treatment groups. In addition, because both of the sample sizes are larger than 30 ($n_T = 31$ and $n_{NT} = 31$), the large sample condition is met. The two required conditions are satisfied.

Calculations (C):

Summary statistics: $n_T = 31$, $\bar{x}_T = 42.81$, $s_T = 9.91$, $n_{NT} = 31$, $\bar{x}_{NT} = 58.67$, and $s_{NT} = 10.42$

Test statistic: $t = \frac{(42.81 - 58.67) - 0}{\sqrt{\frac{(9.91)^2}{31} + \frac{(10.42)^2}{31}}} = -6.14$

Degrees of freedom: $V_T = \frac{s_T^2}{n_T} = \frac{(9.91)^2}{31} = 3.168$,

$V_{NT} = \frac{s_{NT}^2}{n_{NT}} = \frac{(10.42)^2}{31} = 3.50246$ and

$\text{df} = \frac{(V_T + V_{NT})^2}{\frac{V_T^2}{n_T - 1} + \frac{V_{NT}^2}{n_{NT} - 1}} = \frac{(3.168 + 3.50246)^2}{\frac{(3.168)^2}{31-1} + \frac{(3.50246)^2}{31-1}} = 59.8495$, truncated to $\text{df} = 59$.

This is a lower-tailed test (the inequality in H_a is <), so the P-value is the area under the t curve and to the left of –6.14 with 59 degrees of freedom. Therefore, the P-value is $P(t \leq -6.14) \approx 0$.

Communicate Results (C):

Because the P-value of approximately 0 is less than the selected significance level of $\alpha = 0.05$, we reject the null hypothesis. There is convincing evidence that the mean quiz score is lower for students who text than for those students who do not text.

14.17: No. It is not appropriate to use the two-sample z test because the groups are not large enough. We are not told the sizes of the groups, but we know that each is, at most, 81. The sample proportion for the fish oil group is 0.05, and 81(0.05) = 4.05, which is less than 10. So, the conditions for the two-sample z test are not satisfied.

14.19: (a) Using the five-step process (HMC^3):

Hypotheses (H):

We want to determine if there is convincing evidence to claim that there is a significant difference in mean elongation between the squared knot and the Duncan loop for Maxon thread. The treatment means are μ_S, the mean elongation of the square knot in Maxon thread, μ_D, the mean elongation of the Duncan loop in Maxon thread, and $\mu_S - \mu_D$, the difference in treatment means. Test the hypotheses $H_0 : \mu_S - \mu_D = 0$ versus $H_a : \mu_S - \mu_D \neq 0$.

Method (M):

Because the answers to the four key questions are (Q) hypothesis testing, (S) experiment data, (T) one numerical variable, and (N) two treatments, we will consider a two-sample t test for a difference in treatment means. The test statistic is $t = \dfrac{(\bar{x}_S - \bar{x}_D) - (\mu_S - \mu_D)}{\sqrt{\dfrac{s_S^2}{n_S} + \dfrac{s_D^2}{n_D}}}$, and we will use a significance level of $\alpha = 0.05$.

Check (C):

We are told that the tissue specimens were randomly assigned to each type of knot and suture material. In addition, we are told that it is reasonable to regard the elongation distributions as approximately normal. The two required conditions are satisfied.

Calculations (C):

Summary statistics: $n_S = 10$, $\bar{x}_S = 10.0$ mm, $s_S = 0.1$ mm, $n_D = 15$, $\bar{x}_D = 11.0$ mm, and $s_D = 0.3$ mm

Test statistic: $t = \dfrac{(10.0 - 11.0) - 0}{\sqrt{\dfrac{(0.1)^2}{10} + \dfrac{(0.3)^2}{15}}} = -11.952$

Degrees of freedom: $V_S = \dfrac{s_S^2}{n_S} = \dfrac{(0.1)^2}{10} = 0.001$, $V_D = \dfrac{s_D^2}{n_D} = \dfrac{(0.3)^2}{15} = 0.006$

and $\text{df} = \dfrac{(V_S + V_D)^2}{\dfrac{V_S^2}{n_S - 1} + \dfrac{V_D^2}{n_D - 1}} = \dfrac{(0.001 + 0.006)^2}{\dfrac{(0.001)^2}{10-1} + \dfrac{(0.006)^2}{15-1}} = 18.2663$, truncated to $\text{df} = 18$.

This is a two-tailed test (the inequality in H_a is $\neq$), so the P-value is the area under the t curve and to the left of -11.952 and to the right of 11.952 with 18 degrees of freedom. Therefore, using the symmetry of the t curve, the P-value is $2 \cdot P(t \leq -11.952) \approx 0$.

Communicate Results (C):

Because the P-value of approximately 0 is less than the selected significance level of $\alpha = 0.05$, we reject the null hypothesis. There is convincing evidence that the mean elongations for the square knot and the Duncan loop for Maxon thread are not the same.

(b) Using the five-step process (HMC^3):

Hypotheses (H):

We want to determine if there is convincing evidence to claim that there is a significant difference in mean elongation between the squared knot and the Duncan loop for Ticron thread. The treatment means are μ_S, the mean elongation of the square knot in Ticron thread, μ_D, the mean elongation of the Duncan loop in Ticron thread, and $\mu_S - \mu_D$, the difference in treatment means. Test the hypotheses $H_0 : \mu_S - \mu_D = 0$ versus $H_a : \mu_S - \mu_D \neq 0$.

Method (M):

Because the answers to the four key questions are (Q) hypothesis testing, (S) experiment data, (T) one numerical variable, and (N) two treatments, we will consider a two-sample *t* test for a difference in treatment means. The test statistic is $t = \dfrac{(\bar{x}_S - \bar{x}_D) - (\mu_S - \mu_D)}{\sqrt{\dfrac{s_S^2}{n_S} + \dfrac{s_D^2}{n_D}}}$, and we will use a significance level of $\alpha = 0.05$.

Check (C):

We are told that the tissue specimens were randomly assigned to each type of knot and suture material. In addition, we are told that it is reasonable to regard the elongation distributions as approximately normal. The two required conditions are satisfied.

Calculations (C):

Summary statistics: $n_S = 10$, $\bar{x}_S = 2.5$ mm, $s_S = 0.06$ mm, $n_D = 11$, $\bar{x}_D = 10.9$ mm, and $s_D = 0.40$ mm

Test statistic: $t = \dfrac{(2.5 - 10.9) - 0}{\sqrt{\dfrac{(0.06)^2}{10} + \dfrac{(0.40)^2}{11}}} = -68.803$

Degrees of freedom: $V_S = \frac{s_S^2}{n_S} = \frac{(0.06)^2}{10} = 0.00036$,

$V_D = \frac{s_D^2}{n_D} = \frac{(0.40)^2}{11} = 0.014545$ and

$$\text{df} = \frac{(V_S + V_D)^2}{\frac{V_S^2}{n_S - 1} + \frac{V_D^2}{n_D - 1}} = \frac{(0.00036 + 0.014545)^2}{\frac{(0.00036)^2}{10-1} + \frac{(0.014545)^2}{11-1}} = 10.494 \text{, truncated to}$$

$\text{df} = 10$.

This is a two-tailed test (the inequality in H_a is $\neq$), so the P-value is the area under the t curve and to the left of –68.803 and to the right of 68.803 with 10 degrees of freedom. Therefore, using the symmetry of the t curve, the P-value is $2 \cdot P(t \leq -68.803) \approx 0$.

Communicate Results (C):

Because the P-value of approximately 0 is less than the selected significance level of $\alpha = 0.05$, we reject the null hypothesis. There is convincing evidence that the mean elongations for the square knot and the Duncan loop for Ticron thread are not the same.

Section 14.3 Exercise Set 1

14.21: Using the five-step process (EMC^3):

Estimate (E):

We want to estimate the difference in mean food intake (in Kcal) for the two different sleep conditions. The treatment means are μ_4, the mean food intake (in Kcal) for the 4-hour sleep group, μ_8, the mean food intake (in Kcal) for the 8-hour sleep group, and $\mu_4 - \mu_8$, the difference in treatment means.

Method (M):

Because the answers to the four key questions are (Q) estimation, (S) experiment data, (T) one numerical variable, and (N) two treatments, we will consider constructing a two-sample t confidence interval using a confidence level of 95%.

Check (C):

We are told that the men were randomly assigned to one of the two experimental treatment groups (sleep conditions). In addition, because both of the sample sizes are smaller than 30 ($n_4 = 15$ and $n_8 = 15$), we need to be willing to assume that the food intake distributions are approximately normal for each of the two treatments. The boxplots shown below constructed using the sample data are roughly symmetric and since there are no outliers in either sample, the assumption of normality is justified. Therefore, the necessary conditions have been satisfied and it is reasonable to construct a two-sample t confidence interval.

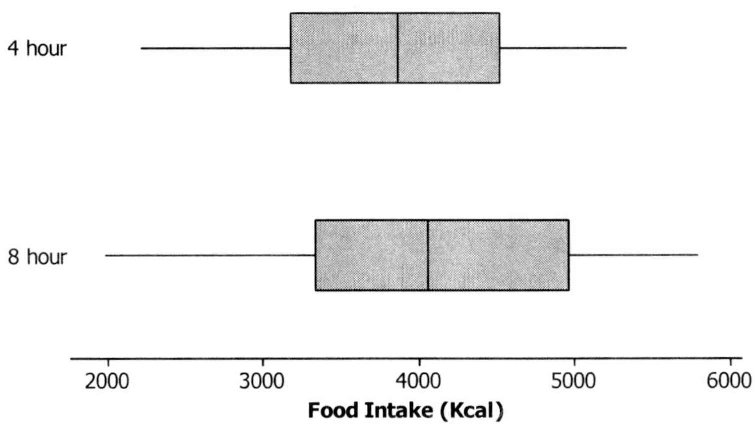

Calculations (C):

Summary statistics: $n_4 = 15$, $\bar{x}_4 = 3,924$ Kcal, $s_4 = 830$ Kcal, $n_8 = 15$, $\bar{x}_8 = 4,069$ Kcal, and $s_8 = 953$ Kcal

Degrees of freedom: $V_4 = \frac{s_4^2}{n_4} = \frac{(830)^2}{15} = 45,926.7$,

$V_8 = \frac{s_8^2}{n_8} = \frac{(953)^2}{15} = 60,547.3$ and

$$\text{df} = \frac{(V_4 + V_8)^2}{\frac{V_4^2}{n_4 - 1} + \frac{V_8^2}{n_8 - 1}} = \frac{(45,926.7 + 60,547.3)^2}{\frac{(45,926.7)^2}{15 - 1} + \frac{(60,547.3)^2}{15 - 1}} = 27.4818.$$

The t critical value found with technology for 95% confidence and using 27.4818 degrees of freedom is 2.05015.

$$(\bar{x}_4 - \bar{x}_8) \pm (t \text{ critical value})\sqrt{\frac{s_4^2}{n_4} + \frac{s_8^2}{n_8}}$$

$$(3{,}924 - 4{,}069) \pm (2.05015)\sqrt{\frac{830^2}{15} + \frac{953^2}{15}}$$

$$-145 \pm (2.05015)(326.303)$$

$$-145 \pm 668.971$$

$$(-813.97, 523.97)$$

Communicate Results (C):

Interpret Confidence Interval: We can be 95% confident that the difference in mean food intake (in Kcal) for the two different sleep conditions is between
–813.97 and 523.97 Kcal. Because 0 is included in the confidence interval, it is possible that there is no difference in the treatment means.

Interpret Confidence Level: The method used to construct this interval estimate is successful in capturing the actual difference in treatment means about 95% of the time.

14.22: (a) If the vertebroplasty group had been compared to a group of patients who did not receive any treatment, and if, for example, the people in the vertebroplasty group experienced a greater pain reduction on average than the people in the "no treatment" group, then it would be impossible to tell whether the observed pain reduction in the vertebroplasty group was caused by the treatment or merely by the subjects' knowledge that some treatment was being applied. By using a placebo group, it is ensured that the subjects in both groups have the knowledge of some "treatment," so that any differences between the pain reduction in the two groups can be attributed to the nature of the vertebroplasty treatment.

(b) Using the five-step process (EMC^3):

Estimate (E):

We want to estimate the difference in mean pain intensity 3 days after the vertebroplasty and placebo treatments. The treatment means are μ_V, the mean pain intensity 3 days after the vertebroplasty treatment, μ_P, the mean pain intensity 3 days after the placebo treatment, and $\mu_V - \mu_P$, the difference in treatment means.

Method (M):

Because the answers to the four key questions are (Q) estimation, (S) experiment data, (T) one numerical variable, and (N) two treatments, we will consider constructing a two-sample *t* confidence interval using a confidence level of 95%.

Check (C):

We are not told whether or not the patients were randomly assigned to one of the two experimental treatment groups (vertebroplasty or placebo). We only know that patients did not know what treatment they received. We will assume that patients were randomly assigned to treatment groups. In addition, because both of the sample sizes are greater than 30 ($n_V = 68$ and $n_P = 63$), the large sample condition has been satisfied. Therefore, the necessary conditions have been satisfied and it is reasonable to construct a two-sample *t* confidence interval.

Calculations (C):

Summary statistics: $n_V = 68$, $\bar{x}_V = 4.2$, $s_V = 2.8$, $n_P = 63$, $\bar{x}_P = 3.9$, and $s_P = 2.9$

Degrees of freedom: $V_V = \frac{s_V^2}{n_V} = \frac{(2.8)^2}{68} = 0.115294$,

$V_P = \frac{s_P^2}{n_P} = \frac{(2.9)^2}{63} = 0.133492$ and

$$\text{df} = \frac{(V_V + V_P)^2}{\frac{V_V^2}{n_V - 1} + \frac{V_P^2}{n_P - 1}} = \frac{(0.115294 + 0.133492)^2}{\frac{(0.115294)^2}{68-1} + \frac{(0.133492)^2}{63-1}} = 127.402.$$

The *t* critical value found with technology for 95% confidence and using 127.402 degrees of freedom is 1.97876.

$$(\overline{x}_V - \overline{x}_P) \pm (t \text{ critical value})\sqrt{\frac{s_V^2}{n_P} + \frac{s_V^2}{n_P}}$$

$$(4.2 - 3.9) \pm (1.97876)\sqrt{\frac{2.8^2}{68} + \frac{2.9^2}{63}}$$

$$0.3 \pm (1.97876)(0.498785)$$

$$0.3 \pm 0.986975$$

$$(-0.687, 1.287)$$

Communicate Results (C):

Interpret Confidence Interval: We can be 95% confident that the difference in mean pain intensity 3 days after treatment for the vertebroplasty treatment and the fake treatment is between −0.687 and 1.287. Because 0 is included in the confidence interval, it is possible that there is no difference in the treatment means.

Interpret Confidence Level: The method used to construct this interval estimate is successful in capturing the actual difference in treatment means about 95% of the time.

14.23: (a) *14 Days after treatment*

Using the five-step process (EMC^3):

Estimate (E):

We want to estimate the difference in mean pain intensity 14 days after the vertebroplasty and placebo treatments. The treatment means are μ_V, the mean pain intensity 14 days after the vertebroplasty treatment, μ_P, the mean pain intensity 14 days after the placebo treatment, and $\mu_V - \mu_P$, the difference in treatment means.

Method (M):

Because the answers to the four key questions are (Q) estimation, (S) experiment data, (T) one numerical variable, and (N) two treatments, we will consider constructing a two-sample *t* confidence interval using a confidence level of 95%.

Check (C):

We are not told whether or not the patients were randomly assigned to one of the two experimental treatment groups (vertebroplasty or placebo). We only know that patients did not know what treatment they received. We will assume that patients were randomly assigned to treatment groups. In addition, because both of the sample sizes are greater than 30 ($n_V = 68$ and $n_P = 63$), the large sample condition has been satisfied. Therefore, the necessary conditions have been satisfied and it is reasonable to construct a two-sample *t* confidence interval.

Calculations (C):

Summary statistics: $n_V = 68$, $\bar{x}_V = 4.3$, $s_V = 2.9$, $n_P = 63$, $\bar{x}_P = 4.5$, and $s_P = 2.8$

Degrees of freedom: $V_V = \dfrac{s_V^2}{n_V} = \dfrac{(2.9)^2}{68} = 0.123676$,

$$V_P = \frac{s_P^2}{n_P} = \frac{(2.8)^2}{63} = 0.12444 \text{ and}$$

$$\text{df} = \frac{(V_V + V_P)^2}{\dfrac{V_V^2}{n_V - 1} + \dfrac{V_P^2}{n_P - 1}} = \frac{(0.123676 + 0.12444)^2}{\dfrac{(0.123676)^2}{68-1} + \dfrac{(0.12444)^2}{63-1}} = 128.774.$$

The *t* critical value found with technology for 95% confidence and using 128.774 degrees of freedom is 1.97856.

$$(\bar{x}_V - \bar{x}_P) \pm (t \text{ critical value})\sqrt{\frac{s_V^2}{n_P} + \frac{s_V^2}{n_P}}$$

$$(4.3 - 4.5) \pm (1.97856)\sqrt{\frac{2.9^2}{68} + \frac{2.8^2}{63}}$$

$$-0.2 \pm (1.97856)(0.498117)$$

$$-0.2 \pm 0.985555$$

$$(-1.186, 0.786)$$

Communicate Results (C):

Interpret Confidence Interval: We can be 95% confident that the difference in mean pain intensity 14 days after treatment for the vertebroplasty treatment and the fake treatment is between −1.186 and 0.786. Because 0 is included in the confidence interval, it is possible that there is no difference in the treatment means.

Interpret Confidence Level: The method used to construct this interval estimate is successful in capturing the actual difference in treatment means about 95% of the time.

One month after treatment

Using the five-step process (EMC^3):

Estimate (E):

We want to estimate the difference in mean pain intensity one month after the vertebroplasty and placebo treatments. The treatment means are μ_V, the mean pain intensity one month after the vertebroplasty treatment, μ_P, the mean pain intensity one month after the placebo treatment, and $\mu_V - \mu_P$, the difference in treatment means.

Method (M):

Because the answers to the four key questions are (Q) estimation, (S) experiment data, (T) one numerical variable, and (N) two treatments, we will consider constructing a two-sample *t* confidence interval using a confidence level of 95%.

Check (C):

We are not told whether or not the patients were randomly assigned to one of the two experimental treatment groups (vertebroplasty or placebo). We only know that patients did not know what treatment they received. We will assume that patients were randomly assigned to treatment groups. In addition, because both of the sample sizes are greater than 30 ($n_V = 68$ and $n_P = 63$), the large sample condition has been satisfied. Therefore, the necessary conditions have been satisfied and it is reasonable to construct a two-sample *t* confidence interval.

Calculations (C):

Summary statistics: $n_V = 68$, $\bar{x}_V = 3.9$, $s_V = 2.9$, $n_P = 63$, $\bar{x}_P = 4.6$, and $s_P = 3.0$

Degrees of freedom: $V_V = \frac{s_V^2}{n_V} = \frac{(2.9)^2}{68} = 0.123676$,

$V_P = \frac{s_P^2}{n_P} = \frac{(3.0)^2}{63} = 0.142857$ and

$$\text{df} = \frac{(V_V + V_P)^2}{\frac{V_V^2}{n_V - 1} + \frac{V_P^2}{n_P - 1}} = \frac{(0.123676 + 0.142857)^2}{\frac{(0.123676)^2}{68-1} + \frac{(0.142857)^2}{63-1}} = 126.387.$$

The t critical value found with technology for 95% confidence and using 126.387 degrees of freedom is 1.97938.

$$(\bar{x}_V - \bar{x}_P) \pm (t \text{ critical value})\sqrt{\frac{s_V^2}{n_P} + \frac{s_V^2}{n_P}}$$

$$(3.9 - 4.6) \pm (1.97938)\sqrt{\frac{2.9^2}{68} + \frac{3.0^2}{63}}$$

$$-0.7 \pm (1.97938)(0.516269)$$

$$-0.7 \pm 1.02189$$

$$(-1.722, 0.322)$$

Communicate Results (C):

Interpret Confidence Interval: We can be 95% confident that the difference in mean pain intensity one month after treatment for the vertebroplasty treatment and the fake treatment is between −1.722 and 0.322. Because 0 is included in the confidence interval, it is possible that there is no difference in the treatment means.

Interpret Confidence Level: The method used to construct this interval estimate is successful in capturing the actual difference in treatment means about 95% of the time.

(b) The fact that all of the intervals contain zero tells us that we do not have convincing evidence of a difference in the mean pain intensity for the vertebroplasty treatment and the fake treatment at any of the three times.

14.24: (a) If people believe that one of the treatments is more effective than the other (for example that the injection is more effective than the spray), it might affect how they interpret and report symptoms.

(b) Using the 5-step process (EMC^3):

Estimate (E): The difference in the proportion of children who would get the flu after being vaccinated with an injection and the proportion of children who would get the flu after being vaccinated with the nasal spray, $p_1 - p_2$, will be estimated, where p_1 is the proportion of children who would get the flu after being vaccinated by the nasal spray and p_2 is the proportion of children who would get the flu after being vaccinated by injection.

Method (M): Because the answers to the four key questions are (Q) estimation, (S) experiment data, (T) one categorical variable, and (N) two treatments, we can construct and interpret a 99% large-sample confidence interval for the difference in proportions of children who get the flu for these two treatments.

Check (C): There is no indication as to how the children were assigned to the injection or nasal spray groups. As such, we must assume that random assignment of children to treatments was used. In addition, $n_1\hat{p}_1 = (4,000)(0.086) = 344$, $n_1(1-\hat{p}_1) = 4,000(1-0.086) = 3,656$, $n_2\hat{p}_2 = (4,000)(0.039) = 156$, and $n_2(1-\hat{p}_2) = 4,000(1-0.039) = 3,844$ are all greater than 10, so the large sample size condition has been satisfied.

Calculations (C):

$$(\hat{p}_1 - \hat{p}_2) \pm (z \text{ critical value})\sqrt{\frac{\hat{p}_1(1-\hat{p}_1)}{n_1} + \frac{\hat{p}_2(1-\hat{p}_2)}{n_2}}$$

$$(0.086 - 0.039) \pm 2.58\sqrt{\frac{0.086(1-0.086)}{4,000} + \frac{0.039(1-0.039)}{4,000}}$$

$$0.047 \pm 2.58(0.005387)$$

$$0.047 \pm 0.013899$$

$$(0.033, 0.061)$$

Communicate Results (C):

Interpret confidence interval: We are 99% confident that the proportion of children who get sick with the flu after being vaccinated with an injection minus the proportion of children who get sick with the flu after being vaccinated with the nasal spray is between 0.033 and 0.061. Since zero is not included in the interval, we have convincing evidence at the 0.01 significance level that the proportions of children who get the flu are different for the two vaccination methods.

Interpret confidence level: The method used to construct this interval estimate is successful in capturing the actual value of the difference in population proportions 99% of the time.

Section 14.3 Additional Exercises

14.27: Using the five-step process (EMC^3):

Estimate (E):

We want to estimate the mean difference in overall course evaluation score. The treatment means are μ_C, the mean course evaluation score for the chocolate treatment group, μ_{NC}, the mean course evaluation score for the no chocolate treatment group, and $\mu_C - \mu_{NC}$, the difference in treatment means.

Method (M):

Because the answers to the four key questions are (Q) estimation, (S) experiment data, (T) one numerical variable, and (N) two treatments, we will consider constructing a two-sample *t* confidence interval using a confidence level of 95%.

Check (C):

We are told that the students were randomly assigned to one of the two experimental treatment groups (chocolate or no chocolate). In addition, because both of the sample sizes are greater than 30 ($n_C = 49$ and $n_{NC} = 49$), the large sample condition has been satisfied. Therefore, the necessary conditions have been satisfied and it is reasonable to construct a two-sample *t* confidence interval.

Calculations (C):

Summary statistics: $n_C = 49$, $\bar{x}_C = 4.07$, $s_C = 0.88$, $n_{NC} = 49$, $\bar{x}_{NC} = 3.85$, and $s_{NC} = 0.89$

Degrees of freedom: $V_C = \dfrac{s_C^2}{n_C} = \dfrac{(0.88)^2}{49} = 0.015804$,

$V_{NC} = \dfrac{s_{NC}^2}{n_{NC}} = \dfrac{(0.89)^2}{49} = 0.016165$ and

$$\text{df} = \frac{(V_C + V_{NC})^2}{\frac{V_C^2}{n_C - 1} + \frac{V_{NC}^2}{n_{NC} - 1}} = \frac{(0.015804 + 0.016165)^2}{\frac{(0.015804)^2}{49-1} + \frac{(0.016165)^2}{49-1}} = 95.9878.$$

The *t* critical value found with technology for 95% confidence and using 95.9878 degrees of freedom is 1.98422.

$$(\bar{x}_C - \bar{x}_{NC}) \pm (t \text{ critical value})\sqrt{\frac{s_C^2}{n_C} + \frac{s_{NC}^2}{n_{NC}}}$$

$$(4.07 - 3.85) \pm (1.98422)\sqrt{\frac{0.88^2}{49} + \frac{0.89^2}{49}}$$

$$0.22 \pm (1.98422)(0.1788)$$

$$0.22 \pm 0.354779$$

$$(-0.135, 0.575)$$

Communicate Results (C):

Interpret Confidence Interval: We can be 95% confident that the difference in mean overall course evaluation is between –0.135 and 0.575. Because 0 is included in the confidence interval, it is possible that there is no difference in the treatment means.

Interpret Confidence Level: The method used to construct this interval estimate is successful in capturing the actual difference in treatment means about 95% of the time.

(b) The interval does not support the statement made in the title. Zero is in the confidence interval, indicating that there may be no real difference between the two treatment means.

14.29: Using the five-step process (EMC^3):

Estimate (E):

We want to estimate the difference in mean quiz score for the texting allowed and texting not allowed treatments. The treatment means are μ_T, the mean quiz score for the texting allowed treatment group, μ_{NT}, the mean quiz score for the no texting allowed treatment group, and $\mu_T - \mu_{NT}$, the difference in treatment means.

Method (M):

Because the answers to the four key questions are (Q) estimation, (S) experiment data, (T) one numerical variable, and (N) two treatments, we will consider constructing a two-sample *t* confidence interval using a confidence level of 90%.

Check (C):

We are told that the students were randomly assigned to one of the two experimental treatment groups (texting allowed or no texting allowed). In addition, because both of the sample sizes are greater than 30 ($n_T = 31$ and $n_{NT} = 31$), the large sample condition has been satisfied. Therefore, the necessary conditions have been satisfied and it is reasonable to construct a two-sample *t* confidence interval.

Calculations (C):

Summary statistics: $n_T = 31$, $\bar{x}_T = 42.81$, $s_T = 9.91$, $n_{NT} = 31$, $\bar{x}_{NT} = 58.67$, and $s_{NT} = 10.42$

Degrees of freedom: $V_T = \frac{s_T^2}{n_T} = \frac{(9.91)^2}{31} = 3.168$,

$$V_{NT} = \frac{s_{NT}^2}{n_{NT}} = \frac{(10.42)^2}{31} = 3.50246 \text{ and}$$

$$\text{df} = \frac{(V_T + V_{NT})^2}{\frac{V_T^2}{n_T - 1} + \frac{V_{NT}^2}{n_{NT} - 1}} = \frac{(3.168 + 3.50246)^2}{\frac{(3.168)^2}{31-1} + \frac{(3.50246)^2}{31-1}} = 59.8495.$$

The t critical value found with technology for 90% confidence and using 59.8495 degrees of freedom is 1.67071.

$$(\bar{x}_T - \bar{x}_{NT}) \pm (t \text{ critical value})\sqrt{\frac{s_T^2}{n_T} + \frac{s_{NT}^2}{n_{NT}}}$$

$$(42.81 - 58.67) \pm (1.67071)\sqrt{\frac{9.91^2}{31} + \frac{10.42^2}{31}}$$

$$-15.86 \pm (1.67071)(2.58272)$$

$$-15.86 \pm 4.31498$$

$$(-20.175, -11.545)$$

Communicate Results (C):

Interpret Confidence Interval: We can be 90% confident that the difference in mean quiz score is between –20.175 and -11.545. Both endpoints of the confidence interval are negative, so you can estimate that the mean quiz score for students in the no texting allowed treatment group is greater than the mean quiz score for students in the texting allowed treatment group by somewhere between 11.545 and 20.175.

Interpret Confidence Level: The method used to construct this interval estimate is successful in capturing the actual difference in treatment means about 90% of the time.

14.31: Statement 3 is correct. The claim is that $\mu_1 > \mu_2$. This means that both endpoints for a confidence interval for $\mu_1 - \mu_2$ would be positive.

Are You Ready To Move On? Chapter 14 Review Exercises

14.33: The difference in mean sleep duration for the caffeine and no-caffeine groups could have occurred by chance just due to the random assignment. This does not mean you should be convinced the treatment means are equal—only that there is not convincing evidence that they are different.

14.35: Using the five-step process (HMC^3):

Hypotheses (H):

We want to determine if there is convincing evidence that the proportion of patients who improve is significantly higher for the experimental treatment than for the standard treatment. The treatment proportions are p_E, the proportion of patients who improve with the experimental treatment, p_S, the proportion of patients who improve with the standard treatment, and $p_E - p_S$, the difference in treatment proportions. Test the hypotheses $H_0 : p_E - p_S = 0$ versus $H_a : p_E - p_S > 0$.

Method (M):

Because the answers to the four key questions are (Q) hypothesis testing, (S) experiment data, (T) one categorical variable, and (N) two treatments, we will consider a two-sample z test for a difference in treatment proportions. The test statistic is $z = \frac{\hat{p}_E - \hat{p}_S}{\sqrt{\frac{\hat{p}_c(1-\hat{p}_c)}{n_E} + \frac{\hat{p}_c(1-\hat{p}_c)}{n_S}}}$, and we will use a significance level of $\alpha = 0.05$.

Check (C):

We are told that the subjects (patients) were randomly assigned to either the experimental treatment group or the standard treatment group. In addition, $n_E\hat{p}_E = 57(0.38) = 21.66$, $n_E(1-\hat{p}_E) = 57(1-0.38) = 35.34$, $n_S\hat{p}_S = 50(0.27) = 13.5$, and $n_S(1-\hat{p}_S) = 50(1-0.27) = 36.5$, which are all larger than 10, so both treatment groups are large enough. The two required conditions are satisfied.

Calculations (C):

Summary statistics: $\hat{p}_E = 0.38$ and $\hat{p}_S = 0.27$

Combined estimate of the common proportion:

$$\hat{p}_c = \frac{n_E \hat{p}_E + n_S \hat{p}_S}{n_E + n_S} = \frac{(57)(0.38) + (50)(0.27)}{57 + 50} = 0.328598$$

Test statistic:

$$z = \frac{0.38 - 0.27}{\sqrt{\frac{0.328598(1-0.328598)}{57} + \frac{0.328598(1-0.328598)}{50}}} = 1.209$$

This is an upper-tailed test (the inequality in H_a is >), so the P-value is the area under the z curve and to the right of 1.209. Therefore, the P-value is $P(z \geq 1.209) = 0.113$.

Communicate Results (C):

Because the P-value of 0.113 is greater than the selected significance level of $\alpha = 0.05$, we fail to reject the null hypothesis. There is not convincing evidence that the proportion of patients who improve is greater for the experimental treatment than for the standard treatment.

14.37: (a) The two treatments are "white-sounding" names and "black-sounding" names.

(b) Using the 5-step process (EMC^3):

Estimate (E): The difference in response proportions for the resumes with "white-sounding" names and with "black-sounding" names, $p_1 - p_2$, will be estimated, where p_1 is the response proportion of resumes with "white-sounding" names and p_2 is the response proportion of resumes with "black-sounding" names.

Method (M): Because the answers to the four key questions are (Q) estimation, (S) experiment data, (T) one categorical variable, and (N) two treatments, we can construct and interpret a 95% large-sample confidence interval for the difference in response proportions for the two treatments.

Check (C): We are told that the job ads were randomly assigned to either the "white-sounding" name group or the "black-sounding" name group. In addition, each of the treatment groups has at least 10 successes (250 resumes received responses in the "white-sounding" name group, and 167 resumes received responses in the "black-sounding" name group) and 10 failures (2,250 resumes did not receive responses in the "white-sounding" name group, and 2,333 resumes did not receive responses in the "black-sounding" name group). The two required conditions are satisfied.

Calculations (C): $\hat{p}_1 = \frac{250}{2,500} = 0.10$ and $\hat{p}_2 = \frac{167}{2,500} = 0.0668$

$$(\hat{p}_1 - \hat{p}_2) \pm (z \text{ critical value})\sqrt{\frac{\hat{p}_1(1-\hat{p}_1)}{n_1} + \frac{\hat{p}_2(1-\hat{p}_2)}{n_2}}$$

$$(0.10 - 0.0668) \pm 1.96\sqrt{\frac{0.10(1-0.10)}{2,500} + \frac{0.0668(1-0.0668)}{2,500}}$$

$$0.0332 \pm 1.96(0.007806)$$

$$0.0332 \pm 0.0153$$

$$(0.018, 0.049)$$

Communicate Results (C):

Interpret confidence interval: We are 95% confident that the difference in the proportions of responses for "white-sounding" names and "black-sounding" names is between 0.018 and 0.049. Both endpoints of the confidence interval are positive, so you can estimate that the proportion of responses for the "white-sounding" names group is greater than the proportion of responses in the "black-sounding" names group by somewhere between 0.018 and 0.049.

Interpret confidence level: The method used to construct this interval estimate is successful in capturing the actual value of the difference in population proportions 95% of the time.

Chapter 15 Learning from Categorical Data

Section 15.1 Exercise Set 1

15.1: (a) With $\text{df} = k-1=4-1=3$, the *P*-value is $P(X^2 \ge 6.4)=0.094$. Because the *P*-value of 0.094 is greater than the significance level of $\alpha=0.05$, we fail to reject H_0. We do not have convincing evidence that the sales are not equally divided among the four colors.

(b) With $\text{df} = k-1=4-1=3$, the *P*-value is $P(X^2 \ge 15.3)=0.002$. Because the *P*-value of 0.002 is less than the significance level of $\alpha=0.01$, we reject H_0. We have convincing evidence that the sales are not equally divided among the four colors.

(c) With $\text{df} = k-1=6-1=5$, the *P*-value is $P(X^2 \ge 13.7)=0.018$. Because the *P*-value of 0.018 is less than the significance level of $\alpha=0.05$, we reject H_0. We have convincing evidence that the sales are not equally divided among the six colors.

15.2: (a) With 3 degrees of freedom, the approximate *P*-value is $P(X^2 \ge 6.62)=0.085$.

(b) With 10 degrees of freedom, the approximate *P*-value is $P(X^2 \ge 16.97)=0.075$.

(c) With 17 degrees of freedom, the approximate *P*-value is $P(X^2 \ge 30.19)=0.025$.

15.3: Using the five-step process (HMC^3):

Hypotheses (H):

We want to determine if there is convincing evidence that the proportions of Twitter users falling into each of the five categories are not all the same.

Define the following population characteristics of interest:

p_1 = proportion of Twitter users in the IS (information sharing) category

p_2 = proportion of Twitter users in the OC (opinions and complaints) category

p_3 = proportion of Twitter users in the RT (random thoughts) category

p_4 = proportion of Twitter users in the ME (me now / what am I doing now) category

p_5 = proportion of Twitter users in the O (other) category

State the appropriate hypotheses:

$H_0: p_1 = p_2 = p_3 = p_4 = p_5 = 0.20$

H_a: At least one of the population proportions is not 0.20

Method (M):

Because the answers to the four key questions are (Q) hypothesis testing, (S) sample data, (T) one categorical variable with more than two categories, and (N) one sample, a chi-square goodness-of-fit test will be considered.

The test statistic for this test is

$$X^2 = \sum_{\text{all categories}} \frac{(\text{observed count} - \text{expected count})^2}{\text{expected count}}$$

When the null hypothesis is true, this statistic has approximately a chi-square distribution with $k - 1$ degrees of freedom.

A significance level of $\alpha = 0.05$ will be used for this test.

Check (C):

The 350 Twitter users in the sample were randomly selected. In addition, all expected counts in each category are all equal to $350(0.20) = 70$, which is greater than 5, so the sample size is large enough. The two conditions necessary to use the chi-square goodness-of-fit test have been satisfied.

Calculations (C):

Test statistic:

$$X^2 = \sum_{\text{all categories}} \frac{(\text{observed count} - \text{expected count})^2}{\text{expected count}}$$

$$= \frac{(51-70)^2}{70} + \frac{(61-70)^2}{70} + \frac{(64-70)^2}{70} + \frac{(101-70)^2}{70} + \frac{(73-70)^2}{70}$$

$$= 5.1571 + 1.1571 + 0.5143 + 13.7286 + 0.1286$$

$$= 20.686$$

Degrees of freedom: $\text{df} = k - 1 = 5 - 1 = 4$

P-value: The P-value is the area under the chi-square curve with 4 degrees of freedom and to the right of 20.686. Therefore, the P-value is equal to $P(X^2 \geq 20.686) \approx 0$.

Communicate Results (C):

Because the P-value of approximately 0 is less than the chosen significance level of $\alpha = 0.05$, we reject H_0. There is convincing evidence that the Twitter category proportions are not all equal.

15.4: (a) Using the five-step process (HMC[3]):

Hypotheses (H):

We want to determine if there is convincing evidence that the proportion of home runs hit is not the same for all five directions.

Define the following population characteristics of interest:

p_1 = proportion of home runs to left field

p_2 = proportion of home runs to left center

p_3 = proportion of home runs to center

p_4 = proportion of home runs to right center

p_5 = proportion of home runs to right field

State the appropriate hypotheses:

$H_0 : p_1 = p_2 = p_3 = p_4 = p_5 = 0.20$

H_a : At least one of the population proportions is not 0.20

Method (M):

Because the answers to the four key questions are (Q) hypothesis testing, (S) sample data, (T) one categorical variable with more than two categories, and (N) one sample, a chi-square goodness-of-fit test will be considered.

The test statistic for this test is

$$X^2 = \sum_{\text{all categories}} \frac{(\text{observed count} - \text{expected count})^2}{\text{expected count}}$$

When the null hypothesis is true, this statistic has approximately a chi-square distribution with $k - 1$ degrees of freedom.

A significance level of $\alpha = 0.05$ will be used for this test.

Check (C):

We are told to assume that this sample of home runs is representative of home runs hit at Yankee Stadium. In addition, all expected counts in each category are all equal to $87(0.20) = 17.4$, which is greater than 5, so the sample size is large enough. The two conditions necessary to use the chi-square goodness-of-fit test have been satisfied.

Calculations (C):

Test statistic:

$$\begin{aligned} X^2 &= \sum_{\text{all categories}} \frac{(\text{observed count} - \text{expected count})^2}{\text{expected count}} \\ &= \frac{(18-17.4)^2}{17.4} + \frac{(10-17.4)^2}{17.4} + \frac{(7-17.4)^2}{17.4} + \frac{(18-17.4)^2}{17.4} + \frac{(34-17.4)^2}{17.4} \\ &= 0.0207 + 3.1471 + 6.2161 + 0.0207 + 15.8368 \\ &= 25.2414 \end{aligned}$$

Degrees of freedom: $\text{df} = k - 1 = 5 - 1 = 4$

P-value: The P-value is the area under the chi-square curve with 4 degrees of freedom and to the right of 25.2414. Therefore, the P-value is equal to $P(X^2 \geq 25.2414) \approx 0$.

Communicate Results (C):

Because the P-value of approximately 0 is less than the chosen significance level of $\alpha = 0.05$, we reject H_0. There is convincing evidence that the proportions of home runs hit are not the same for all five directions.

(b) For home runs going to left center and center field the observed counts are significantly lower than the numbers that would have been expected if the proportion of home runs hit was the same for all five directions, while for right field the observed count is much high than the number that would have been expected.

15.5: Using the five-step process (HMC^3):

Hypotheses (H):

We want to determine if there is convincing evidence to conclude that one or more of these three age groups buys a disproportionate share of lottery tickets.

Define the following population characteristics of interest:

p_1 = proportion of lottery ticket purchasers age 18-34

p_2 = proportion of lottery ticket purchasers age 35-64

p_3 = proportion of lottery ticket purchasers age 65 and over

State the appropriate hypotheses:

$H_0 : p_1 = 0.35, p_2 = 0.51, p_3 = 0.14$

H_a : At least one of the population proportions is not as stated in H_0

Method (M):

Because the answers to the four key questions are (Q) hypothesis testing, (S) sample data, (T) one categorical variable with more than two categories, and (N) one sample, a chi-square goodness-of-fit test will be considered.

The test statistic for this test is

$$X^2 = \sum_{\text{all categories}} \frac{(\text{observed count} - \text{expected count})^2}{\text{expected count}}$$

When the null hypothesis is true, this statistic has approximately a chi-square distribution with $k - 1$ degrees of freedom.

A significance level of $\alpha = 0.05$ will be used for this test.

Check (C):

We are told to suppose that these data resulted from a random sample of 200 lottery ticket purchasers. The expected counts in each category are $200(0.35) = 70$ in the 18-34 age category, $200(0.51) = 102$ in the 35-64 age category, and $200(0.14) = 28$ in the 65 and over age category, which are all greater than 5, so the sample size is large enough. The two conditions necessary to use the chi-square goodness-of-fit test have been satisfied.

Calculations (C):

Test statistic:

$$\begin{aligned} X^2 &= \sum_{\text{all categories}} \frac{(\text{observed count} - \text{expected count})^2}{\text{expected count}} \\ &= \frac{(36-70)^2}{70} + \frac{(130-102)^2}{102} + \frac{(34-28)^2}{28} \\ &= 16.5143 + 7.6863 + 1.2857 \\ &= 25.4863 \end{aligned}$$

Degrees of freedom: $\text{df} = k - 1 = 3 - 1 = 2$

P-value: The *P*-value is the area under the chi-square curve with 2 degrees of freedom and to the right of 25.4863. Therefore, the *P*-value is equal to $P(X^2 \geq 25.4863) \approx 0$.

Communicate Results (C):

Because the P-value of approximately 0 is less than the chosen significance level of $\alpha = 0.05$, we reject H_0. There is convincing evidence that one or more of the age groups buys a disproportionate share of lottery tickets.

Section 15.1 Additional Exercises

15.11: (a) With 13 degrees of freedom, the approximate P-value is $P(X^2 \geq 34.52) = 0.001$.

(b) With 16 degrees of freedom, the approximate P-value is $P(X^2 \geq 39.25) = 0.001$.

(c) With 19 degrees of freedom, the approximate P-value is $P(X^2 \geq 26.00) = 0.130$.

15.13: Because the P-value of less than 0.001 is smaller than any reasonable significance level, we would reject H_0. There is convincing evidence that the response proportions are not each 0.5.

15.15: Using the five-step process (HMC^3):

Hypotheses (H):

We want to determine if there is convincing evidence to conclude that the data from this experiment are not consistent with Mendel's laws.

Define the following population characteristics of interest:

p_1 = proportion of tall cut-leaf phenotypes

p_2 = proportion of tall potato-leaf phenotypes

p_3 = proportion of dwarf-cut leaf phenotypes

p_4 = proportion of dwarf potato-leaf phenotypes

State the appropriate hypotheses:

$$H_0 : p_1 = \frac{9}{16}, p_2 = \frac{3}{16}, p_3 = \frac{3}{16}, p_4 = \frac{1}{16}$$

H_a : At least one of the population proportions is not as stated in H_0

Method (M):

Because the answers to the four key questions are (Q) hypothesis testing, (S) sample data, (T) one categorical variable with more than two categories, and (N) one sample, a chi-square goodness-of-fit test will be considered.

The test statistic for this test is

$$X^2 = \sum_{\text{all categories}} \frac{(\text{observed count} - \text{expected count})^2}{\text{expected count}}$$

When the null hypothesis is true, this statistic has approximately a chi-square distribution with $k - 1$ degrees of freedom.

A significance level of $\alpha = 0.01$ will be used for this test.

Check (C):

We are not told how the sample was selected or if it is representative of all such tomato plants, so must assume that one of these conditions is satisfied. The expected counts in each category are $1{,}611\left(\frac{9}{16}\right) = 906.188$ in the tall cut-leaf phenotype category, $1{,}611\left(\frac{3}{16}\right) = 302.063$ in both the tall potato-leaf and dwarf cut-leaf phenotype categories, and $1{,}611\left(\frac{1}{16}\right) = 100.688$ in the dwarf potato-leaf phenotype category, which are all greater than 5, so the sample size is large enough. The two conditions necessary to use the chi-square goodness-of-fit test have been satisfied.

Calculations (C):

Test statistic:

$$X^2 = \sum_{\text{all categories}} \frac{(\text{observed count} - \text{expected count})^2}{\text{expected count}}$$

$$= \frac{(926-906.188)^2}{906.188} + \frac{(288-302.063)^2}{302.063} + \frac{(293-302.063)^2}{302.063} + \frac{(104-100.688)^2}{100.688}$$
$$= 0.433172 + 0.654679 + 0.271894 + 0.108977$$
$$= 1.46872$$

Degrees of freedom: $\text{df} = k-1 = 4-1 = 3$

P-value: The P-value is the area under the chi-square curve with 3 degrees of freedom and to the right of 1.46872. Therefore, the P-value is equal to $P(X^2 \geq 1.46872) = 0.690$.

Communicate Results (C):

Because the P-value of 0.690 is much greater than the chosen significance level of $\alpha = 0.01$, we fail to reject H_0. There is not convincing evidence to conclude that the data from this experiment are not consistent with Mendel's laws.

15.17: Using the five-step process (HMC³):

Hypotheses (H):

We want to determine if there is convincing evidence that the 1-day old bobwhites have a color preference.

Define the following population characteristics of interest:

p_1 = proportion of bobwhites whose first peck is blue

p_2 = proportion of bobwhites whose first peck is green

p_3 = proportion of bobwhites whose first peck is yellow

p_4 = proportion of bobwhites whose first peck is red

State the appropriate hypotheses:

$H_0: p_1 = p_2 = p_3 = p_4 = 0.25$

H_a: At least one of the population proportions is not 0.25

Method (M):

Because the answers to the four key questions are (Q) hypothesis testing, (S) sample data, (T) one categorical variable with more than two categories, and (N) one sample, a chi-square goodness-of-fit test will be considered.

The test statistic for this test is

$$X^2 = \sum_{\text{all categories}} \frac{(\text{observed count} - \text{expected count})^2}{\text{expected count}}$$

When the null hypothesis is true, this statistic has approximately a chi-square distribution with $k - 1$ degrees of freedom.

A significance level of $\alpha = 0.01$ will be used for this test.

Check (C):

We are not told how the 1-day old bobwhites in the study were selected, so we must assume they were randomly selected or that they form a representative sample. In addition, all expected counts in each category are all equal to $33(0.25) = 8.25$, which is greater than 5, so the sample size is large enough. The two conditions necessary to use the chi-square goodness-of-fit test have been satisfied.

Calculations (C):

Test statistic:

$$\begin{aligned} X^2 &= \sum_{\text{all categories}} \frac{(\text{observed count} - \text{expected count})^2}{\text{expected count}} \\ &= \frac{(16-8.25)^2}{8.25} + \frac{(8-8.25)^2}{8.25} + \frac{(6-8.25)^2}{8.25} + \frac{(3-8.25)^2}{8.25} \\ &= 7.28030 + 0.00758 + 0.61364 + 3.34091 \\ &= 11.2424 \end{aligned}$$

Degrees of freedom: $\text{df} = k - 1 = 4 - 1 = 3$

P-value: The *P*-value is the area under the chi-square curve with 3 degrees of freedom and to the right of 11.2424. Therefore, the *P*-value is equal to $P(X^2 \geq 11.2424) = 0.0105$.

Communicate Results (C):

Because the *P*-value of 0.0105 is greater than the chosen significance level of $\alpha = 0.01$, we fail to reject H_0. There is not convincing evidence that there is a color preference.

Section 15.2 Exercise Set 1

15.19: Using the five-step process (HMC^3):

Hypotheses (H):

We want to determine if there is convincing evidence that the proportion falling in the three credit card response categories is not the same for all three years.

The hypotheses of interest are:

H_0: The proportions falling into the three credit card response categories are the same for all three years

H_a: The proportions falling into the three credit card response categories are not the same for all three years

Method (M):

The purpose of this study is to compare three populations (the three years 2004, 2005, and 2006) on the basis of one categorical variable (credit card payment response), so a chi-square test of homogeneity will be considered.

The test statistic for this test is

$$X^2 = \sum_{\text{all categories}} \frac{(\text{observed count} - \text{expected count})^2}{\text{expected count}}$$

When the null hypothesis is true, this statistic has approximately a chi-square distribution with (number of rows – 1)(number of columns – 1) degrees of freedom.

A significance level of $\alpha = 0.05$ will be used for this test.

Check (C):

We are told that in each year (2004, 2005, and 2006) a random sample of people age 18 or older was selected. The table below shows observed counts and, in parentheses, expected counts. Note that all expected counts are at least 5.

	2004	2005	2006
Definitely Will/Probably Will	40 (43.33)	50 (43.33)	40 (43.33)
Might/Might Not/Probably Not	180 (176.67)	190 (176.67)	160 (176.67)
Definitely Will Not	781 (781)	761 (781)	801 (781)

Because the samples were independent random samples and all expected counts are at least 5, the chi-square test of homogeneity is appropriate.

Calculations (C):

Test statistic:

$$X^2 = \sum_{\text{all categories}} \frac{(\text{observed count} - \text{expected count})^2}{\text{expected count}}$$

$$= \frac{(40-43.3333)^2}{43.3333} + \frac{(50-43.3333)^2}{43.3333} + \cdots + \frac{(761-781)^2}{781} + \frac{(801-781)^2}{781}$$

$$= 0.256 + 1.026 + \cdots + 0.512 + 0.512$$

$$= 5.204$$

Degrees of freedom:
$\text{df} = (\text{number of rows} - 1)(\text{number of columns} - 1) = (2)(2) = 4$

P-value: The *P*-value is the area under the chi-square curve with 4 degrees of freedom and to the right of 5.204. Therefore, the *P*-value is equal to $P(X^2 \geq 5.204) = 0.267$.

Communicate Results (C):

Because the *P*-value of 0.267 is greater than the specified significance level of $\alpha = 0.05$, we fail to reject H_0. There is not convincing evidence that the proportions falling into the three credit response categories are not all the same for all three years.

15.20: (a) Using the five-step process (HMC^3):

Hypotheses (H):

We want to determine if there is convincing evidence that the proportions in the two donation categories are not the same for all three types of requests.

The hypotheses of interest are:

H_0: donation proportions are the same for all three gift types

H_a: donation proportions are not the same for all three gift types

Method (M):

The purpose of this study is to compare three treatments (the three gift categories) on the basis of one categorical variable (donation result), so a chi-square test of homogeneity will be considered.

The test statistic for this test is

$$X^2 = \sum_{\text{all categories}} \frac{(\text{observed count} - \text{expected count})^2}{\text{expected count}}$$

When the null hypothesis is true, this statistic has approximately a chi-square distribution with (number of rows – 1)(number of columns – 1) degrees of freedom.

A significance level of $\alpha = 0.01$ will be used for this test.

Check (C):

We are told that potential donors who were to receive the letter were assigned, at random, to one of the three groups (no gift, small gift, large gift). The table below shows observed counts and, in parentheses, expected counts. Note that all expected counts are at least 5.

	Donation	No Donation
No Gift	397 (514.51)	2,865 (2,747.49)
Small Gift	465 (510.57)	2,772 (2,726.43)
Large Gift	691 (527.92)	2,656 (2,819.08)

Because the treatments were randomly assigned and all expected counts are at least 5, the chi-square test of homogeneity is appropriate.

Calculations (C):

Test statistic:

$$X^2 = \sum_{\text{all categories}} \frac{(\text{observed count} - \text{expected count})^2}{\text{expected count}}$$

$$= \frac{(397-514.51)^2}{514.51} + \frac{(2{,}865-2{,}747.49)^2}{2{,}747.49} + \cdots + \frac{(691-527.92)^2}{527.92} + \frac{(2{,}656-2{,}819.08)^2}{2{,}819.08}$$

$$= 26.839 + 5.026 + \cdots + 50.378 + 9.434$$

$$= 96.506$$

Degrees of freedom:
$\text{df} = (\text{number of rows} - 1)(\text{number of columns} - 1) = (2)(1) = 2$

P-value: The *P*-value is the area under the chi-square curve with 2 degrees of freedom and to the right of 96.506. Therefore, the *P*-value is equal to $P(X^2 \geq 96.506) \approx 0$.

Communicate Results (C):

Because the *P*-value of approximately 0 is smaller than the specified significance level of $\alpha = 0.01$, we reject H_0. There is convincing evidence that the donation proportions are not the same for all three gift types.

(b) The result of part (a) tells us that the level of the gift seems to make a difference. Looking at the data given, 12% of those receiving no gift made a donation, 14% of those receiving a small gift made a donation, and 21% of those receiving a large gift made a donation. (These percentages can be compared to 16% making donations among the expected counts.) So, it seems that the most effective strategy is to include a large gift, with the small gift making very little difference compared to no gift at all.

15.21: (a) Using the five-step process (HMC^3):

Hypotheses (H):

We want to determine if there is convincing evidence that field of study and smoking status are not independent.

The hypotheses of interest are:

H_0: field of study and smoking status are independent

H_a: field of study and smoking status are not independent

Method (M):

The purpose of this study is to compare two categorical variables (field of study and smoking status) from one population (college students at the University of Minnesota), so a chi-square test of independence will be considered.

The test statistic for this test is

$$X^2 = \sum_{\text{all categories}} \frac{(\text{observed count} - \text{expected count})^2}{\text{expected count}}$$

When the null hypothesis is true, this statistic has approximately a chi-square distribution with (number of rows – 1)(number of columns – 1) degrees of freedom.

A significance level of $\alpha = 0.01$ will be used for this test.

Check (C):

We are told that a large random sample of students at the University of Minnesota was taken. The Minitab output in the textbook shows that all expected counts are at least 5. Because the sample was randomly selected and all expected counts are at least 5, the chi-square test of independence is appropriate.

Calculations (C):

Test statistic:

$$\begin{aligned} X^2 &= \sum_{\text{all categories}} \frac{(\text{observed count} - \text{expected count})^2}{\text{expected count}} \\ &= \frac{(176-189.23)^2}{189.23} + \frac{(489-475.77)^2}{475.77} + \cdots + \frac{(134-112.12)^2}{112.12} + \frac{(260-281.88)^2}{281.88} \\ &= 0.925 + 0.368 + \cdots + 4.272 + 1.699 \\ &= 90.853 \end{aligned}$$

Degrees of freedom:
$\text{df} = (\text{number of rows} - 1)(\text{number of columns} - 1) = (8)(1) = 8$

P-value: The P-value is the area under the chi-square curve with 8 degrees of freedom and to the right of 90.853. Therefore, the P-value is equal to $P(X^2 \geq 90.853) \approx 0$.

Communicate Results (C):

Because the P-value of approximately 0 is smaller than the specified significance level of $\alpha = 0.01$, we reject H_0. There is convincing evidence that the variables field of study and smoking status are not independent.

(b) The particularly high contributions to the chi-square statistic (in order of importance) come from the field of communication and languages, in which there was a disproportionately high number of smokers; from the field of mathematics, engineering, and sciences, in which there was a disproportionately low number of smokers; and from the field of social science and human services, in which there was a disproportionately high number of smokers.

15.22: Using the five-step process (HMC^3):

Hypotheses (H):

We want to determine if there is convincing evidence that locus of control and compulsive buyer behavior are associated (not independent).

The hypotheses of interest are:

H_0: locus of control and compulsive buyer behavior are independent

H_a: locus of control and compulsive buyer behavior are not independent

Method (M):

The purpose of this study is to compare two categorical variables (locus of control and compulsive buyer behavior) from one population (college students at Midwestern public universities), so a chi-square test of association (independence) will be considered.

The test statistic for this test is

$$X^2 = \sum_{\text{all categories}} \frac{(\text{observed count} - \text{expected count})^2}{\text{expected count}}$$

When the null hypothesis is true, this statistic has approximately a chi-square distribution with (number of rows – 1)(number of columns – 1) degrees of freedom.

A significance level of $\alpha = 0.01$ will be used for this test.

Check (C):

We are told to assume that the sample was representative of college students at Midwestern public universities. The table below shows observed counts and, in parentheses, expected counts. Note that all expected counts are at least 5.

		Locus of Control	
		Internal	External
Compulsive Buyer?	Yes	3 (7.42)	14 (9.58)
	No	52 (47.58)	57 (61.42)

Because the sample was representative of the population and all expected counts are at least 5, the chi-square test of association (independence) is appropriate.

Calculations (C):

Test statistic:

$$\begin{aligned} X^2 &= \sum_{\text{all categories}} \frac{(\text{observed count} - \text{expected count})^2}{\text{expected count}} \\ &= \frac{(3-7.42)^2}{7.42} + \frac{(14-9.58)^2}{9.58} + \frac{(52-47.58)^2}{47.58} + \frac{(57-61.42)^2}{61.42} \\ &= 2.633 + 2.040 + 0.411 + 0.318 \\ &= 5.402 \end{aligned}$$

Degrees of freedom:

$\text{df} = (\text{number of rows} - 1)(\text{number of columns} - 1) = (1)(1) = 1$

P-value: The P-value is the area under the chi-square curve with 1 degree of freedom and to the right of 5.402. Therefore, the P-value is equal to $P(X^2 \geq 5.402) = 0.020$.

Communicate Results (C):

Because the P-value of 0.020 is greater than the specified significance level of $\alpha = 0.01$, we fail to reject H_0. There is not convincing evidence that there is an association between locus of control and compulsive buyer behavior.

Section 15.2 Additional Exercises

15.27: Answers may vary; one example follows. A political science researcher might be interested in seeing if there is evidence that the distribution of self-described political philosophy (liberal, moderate, conservative) among registered voters has changed over the years 2008, 2009, 2010, and 2011. Therefore, the populations that would be sampled are registered voters in each of the years 2008, 2009, 2010, and 2011. The variable that would be recorded is political philosophy (liberal, moderate, or conservative).

15.29: In a chi-square goodness-of-fit test, one population is compared to fixed category proportions. In a chi-square test of homogeneity, two or more populations are compared.

15.31: Using the five-step process (HMC^3):

Hypotheses (H):

We want to determine if there is convincing evidence that the use of ART and whether or not a baby is premature are associated (not independent).

The hypotheses of interest are:

H_0: There is no association between use of ART and whether or not a baby is premature

H_a: There is an association between use of ART and whether or not a baby is premature

Method (M):

The purpose of this study is to compare two categorical variables (use of ART and whether or not a baby is premature) from one population (births), so a chi-square test of association (independence) will be considered.

The test statistic for this test is

$$X^2 = \sum_{\text{all categories}} \frac{(\text{observed count} - \text{expected count})^2}{\text{expected count}}$$

When the null hypothesis is true, this statistic has approximately a chi-square distribution with (number of rows – 1)(number of columns – 1) degrees of freedom.

A significance level of $\alpha = 0.01$ will be used for this test.

Check (C):

We are told that a large random sample of babies was taken. The table below shows observed counts and, in parentheses, expected counts. Note that all expected counts are at least 5.

	Conceived using ART	Conceived Naturally
Premature	154 (43.33)	2,405 (2,515.67)
Not Premature	212 (322.67)	18.843 (18,732.33)

Because the sample of births was randomly selected and all expected counts are at least 5, the chi-square test of association (independence) is appropriate.

Calculations (C):

Test statistic:

$$X^2 = \sum_{\text{all categories}} \frac{(\text{observed count} - \text{expected count})^2}{\text{expected count}}$$

$$= \frac{(154-43.33)^2}{43.33} + \frac{(2,405-2,515.67)^2}{2,515.67} + \frac{(212-322.67)^2}{322.67} + \frac{(18,843-18,732.33)^2}{18,732.33}$$

$$= 282.632 + 4.868 + 37.956 + 0.654$$

$$= 326.111$$

Degrees of freedom:
$\text{df} = (\text{number of rows} - 1)(\text{number of columns} - 1) = (1)(1) = 1$

P-value: The P-value is the area under the chi-square curve with 1 degree of freedom and to the right of 326.111. Therefore, the P-value is equal to $P(X^2 \geq 326.111) \approx 0$.

Communicate Results (C):

Because the P-value of approximately 0 is less than the specified significance level of $\alpha = 0.01$, we reject H_0. There is convincing evidence that there is an association between use of ART and whether or not a baby is premature.

15.33: Using the five-step process (HMC[3]):

Hypotheses (H):

We want to determine if there is convincing evidence that the proportion who believe that the story describes a rape differs for the three different photo groups.

The hypotheses of interest are:

H_0: The proportions who believe that the story described a rape are the same for the three photo treatments

H_a: The proportions who believe that the story described a rape are not the same for the three photo treatments

Method (M):

The purpose of this study is to compare one categorical variable (situation described was one of rape) from three populations (picture type – provocative, conservative, or no picture), so a chi-square test of homogeneity will be considered.

The test statistic for this test is

$$X^2 = \sum_{\text{all categories}} \frac{(\text{observed count} - \text{expected count})^2}{\text{expected count}}$$

When the null hypothesis is true, this statistic has approximately a chi-square distribution with (number of rows – 1)(number of columns – 1) degrees of freedom.

A significance level of $\alpha = 0.01$ will be used for this test.

Check (C):

We are not told that students were randomly assigned to one of the treatments (different type of photo), but we will assume that is the case. The table below shows observed counts and, in parentheses, expected counts. Note that all expected counts are at least 5.

	Picture		
Response	Provocative	Conservative	No Photo
Rape	80 (99.58)	104 (90.95)	92 (85.47)
Not Rape	47 (27.42)	12 (25.05)	17 (23.53)

Because we are assuming the pictures were randomly assigned to students and all expected counts are at least 5, the chi-square test of homogeneity is appropriate.

Calculations (C):

Test statistic:

$$X^2 = \sum_{\text{all categories}} \frac{(\text{observed count} - \text{expected count})^2}{\text{expected count}}$$

$$= \frac{(80-99.58)^2}{99.58} + \frac{(104-90.95)^2}{90.95} + \frac{(92-85.47)^2}{85.47}$$

$$+ \frac{(47-27.42)^2}{27.42} + \frac{(12-25.05)^2}{25.05} + \frac{(17-23.53)^2}{23.53}$$

$$= 3.850 + 1.871 + 0.500 + 13.981 + 6.795 + 1.814$$

$$= 28.810$$

Degrees of freedom:

$$\text{df} = (\text{number of rows} - 1)(\text{number of columns} - 1) = (2)(1) = 2$$

P-value: The P-value is the area under the chi-square curve with 2 degrees of freedom and to the right of 28.810. Therefore, the P-value is equal to $P(X^2 \geq 28.810) \approx 0$.

Communicate Results (C):

Because the P-value of approximately 0 is less than the specified significance level of $\alpha = 0.01$, we reject H_0. There is convincing evidence that the proportions who believe that the story described a rape are not the same for the three photo treatments.

15.35: Using the five-step process (HMC^3):

Hypotheses (H):

We want to determine if there is convincing evidence that city of residence and vehicle type are associated.

The hypotheses of interest are:

H_0: There is no association between city of residence and vehicle type

H_a: There is an association between city of residence and vehicle type

Method (M):

The purpose of this study is to compare two categorical variables (city of residence and vehicle type) from one population (individuals in the San Francisco bay area), so a chi-square test of association will be considered.

The test statistic for this test is

$$X^2 = \sum_{\text{all categories}} \frac{(\text{observed count} - \text{expected count})^2}{\text{expected count}}$$

When the null hypothesis is true, this statistic has approximately a chi-square distribution with (number of rows – 1)(number of columns – 1) degrees of freedom.

A significance level of $\alpha = 0.05$ will be used for this test.

Check (C):

We are told to assume that the sample is a random sample of San Francisco Bay area residents. The table below shows observed counts and, in parentheses, expected counts. Note that all expected counts are at least 5.

Vehicle Type	City: Concord	Pleasant Hills	North San Francisco
Small	68 (89.06)	83 (107.02)	221 (175.92)
Compact	63 (56.74)	68 (68.18)	106 (112.08)
Midsize	88 (84.51)	123 (101.55)	142 (166.94)
Large	24 (12.69)	18 (15.25)	11 (25.06)

Because we were told to assume the sample was randomly selected and all expected counts are at least 5, the chi-square test of association is appropriate.

Calculations (C):

Test statistic:

$$X^2 = \sum_{\text{all categories}} \frac{(\text{observed count} - \text{expected count})^2}{\text{expected count}}$$

$$= \frac{(68-89.06)^2}{89.06} + \frac{(83-107.02)^2}{107.02} + \cdots + \frac{(18-15.25)^2}{15.25} + \frac{(11-25.06)^2}{25.06}$$

$$= 4.980 + 5.391 + \cdots + 0.497 + 7.892$$

$$= 49.813$$

Degrees of freedom:

$\text{df} = (\text{number of rows} - 1)(\text{number of columns} - 1) = (3)(2) = 6$

P-value: The *P*-value is the area under the chi-square curve with 6 degrees of freedom and to the right of 49.813. Therefore, the *P*-value is equal to $P(X^2 \geq 49.813) \approx 0$.

Communicate Results (C):

Because the P-value of approximately 0 is less than the specified significance level of $\alpha = 0.05$, we reject H_0. There is convincing evidence that there is an association between city of residence and vehicle type.

15.37: Using the five-step process (HMC^3):

Hypotheses (H):

We want to determine if there is convincing evidence that age of children and parental response are associated.

The hypotheses of interest are:

H_0: There is no association between age of children and parental response

H_a: There is an association between age of children and parental response

Method (M):

The purpose of this study is to compare two categorical variables (age of children and parental response) from one population (parents of preteen and teenage children), so a chi-square test of association will be considered.

The test statistic for this test is

$$X^2 = \sum_{\text{all categories}} \frac{(\text{observed count} - \text{expected count})^2}{\text{expected count}}$$

When the null hypothesis is true, this statistic has approximately a chi-square distribution with (number of rows – 1)(number of columns – 1) degrees of freedom.

A significance level of $\alpha = 0.05$ will be used for this test.

Check (C):

We are only told that a telephone survey of parents with preteen and teenage children was taken, and not how the respondents were selected. We will assume the sample was selected randomly. The table below shows observed counts and, in parentheses, expected counts. Note that all expected counts are at least 5.

Response	Age of Children: Preteen	Teen
Very Effective	126 (139.07)	149 (135.93)
Somewhat Effective	44 (42.99)	41 (42.01)
Not at all Effective or Don't Know	51 (38.94)	26 (38.06)

Because we are assuming the sample was randomly selected and all expected counts are at least 5, the chi-square test of association is appropriate.

Calculations (C):

Test statistic:

$$X^2 = \sum_{\text{all categories}} \frac{(\text{observed count} - \text{expected count})^2}{\text{expected count}}$$

$$= \frac{(126-139.07)^2}{139.07} + \frac{(149-135.93)^2}{135.93} + \cdots + \frac{(51-38.94)^2}{38.94} + \frac{(26-38.06)^2}{38.06}$$

$$= 1.229 + 1.257 + \cdots + 3.735 + 3.821$$

$$= 10.091$$

Degrees of freedom:

$$\text{df} = (\text{number of rows} - 1)(\text{number of columns} - 1) = (2)(1) = 2$$

P-value: The P-value is the area under the chi-square curve with 2 degrees of freedom and to the right of 10.091. Therefore, the P-value is equal to $P(X^2 \geq 10.091) = 0.006$.

Communicate Results (C):

Because the P-value of 0.006 is less than the specified significance level of $\alpha = 0.05$, we reject H_0. There is convincing evidence that there is an association between age of children and parental response.

15.39: (a) Since the study was conducted using separate random samples of male and female inmates, this is a test for homogeneity.

(b) Using the five-step process (HMC^3):

Hypotheses (H):

We want to determine if there is convincing evidence that the proportions of types of crime are not all the same for male and female inmates.

The hypotheses of interest are:

H_0: the proportions falling into the crime categories are the same for male and female inmates

H_a: The proportions falling into the crime categories are not the same for male and female inmates

Method (M):

The purpose of this study is to compare genders on the basis of one categorical variable (type of crime), so a chi-square test of homogeneity will be considered.

The test statistic for this test is

$$X^2 = \sum_{\text{all categories}} \frac{(\text{observed count} - \text{expected count})^2}{\text{expected count}}$$

When the null hypothesis is true, this statistic has approximately a chi-square distribution with (number of rows – 1)(number of columns – 1) degrees of freedom.

A significance level of $\alpha = 0.05$ will be used for this test.

Check (C):

We are told that two random samples (500 male inmates and 500 female inmates) were taken. The table below shows observed counts and, in parentheses, expected counts. Note that all expected counts are at least 5.

	Gender	
Type of Crime	Male	Female
Violent	117 (91.5)	66 (91.5)
Property	150(155)	160 (155)
Drug	109 (138.5)	168 (138.5)
Public-order	124 (115)	106 (115)

Because the inmates were randomly selected and all expected counts are at least 5, the chi-square test of homogeneity is appropriate.

Calculations (C):

Test statistic:

$$\begin{aligned}X^2 &= \sum_{\text{all categories}} \frac{(\text{observed count} - \text{expected count})^2}{\text{expected count}} \\ &= \frac{(117-91.5)^2}{91.5} + \frac{(66-91.5)^2}{91.5} + \cdots + \frac{(124-115)^2}{115} + \frac{(106-115)^2}{115} \\ &= 7.107 + 7.107 + \cdots + 0.704 + 0.704 \\ &= 28.511\end{aligned}$$

Degrees of freedom:
$\text{df} = (\text{number of rows} - 1)(\text{number of columns} - 1) = (3)(1) = 3$

P-value: The P-value is the area under the chi-square curve with 3 degrees of freedom and to the right of 28.511. Therefore, the P-value is equal to $P(X^2 \geq 28.511) \approx 0$.

Communicate Results (C):

Because the P-value of approximately 0 is smaller than the specified significance level of $\alpha = 0.05$, we reject H_0. There is convincing evidence that the proportions falling into the crime categories are not all the same for male and female inmates.

Are You Ready To Move On? Chapter 15 Review Exercises

15.41: Using the five-step process (HMC^3):

Hypotheses (H):

We want to determine if there is convincing evidence that fatal bicycle accidents are not equally likely to occur in each of the 3-hour time periods.

Define the following population characteristics of interest:

p_1 = proportion of fatal bicycle accidents from Midnight to 3 A.M.

p_2 = proportion of fatal bicycle accidents from 3 A.M.to 6 A.M.

p_3 = proportion of fatal bicycle accidents from 6 A.M. to 9 A.M.

p_4 = proportion of fatal bicycle accidents from 9 A.M. to Noon

p_5 = proportion of fatal bicycle accidents from Noon to 3 P.M.

p_6 = proportion of fatal bicycle accidents from 3 P.M. to 6 P.M.

p_7 = proportion of fatal bicycle accidents from 6 P.M. to 9 P.M.

p_8 = proportion of fatal bicycle accidents from 9 P.M. to Midnight

State the appropriate hypotheses:

$$H_0 : p_1 = p_2 = p_3 = p_4 = p_5 = p_6 = p_7 = p_8 = \frac{1}{8}$$

H_a : At least one of the population proportions is not $\frac{1}{8}$

Method (M):

Because the answers to the four key questions are (Q) hypothesis testing, (S) sample data, (T) one categorical variable with more than two categories, and (N) one sample, a chi-square goodness-of-fit test will be considered.

The test statistic for this test is

$$X^2 = \sum_{\text{all categories}} \frac{(\text{observed count} - \text{expected count})^2}{\text{expected count}}$$

When the null hypothesis is true, this statistic has approximately a chi-square distribution with $k - 1$ degrees of freedom.

A significance level of $\alpha = 0.05$ will be used for this test.

Check (C):

We are told to assume these 715 bicycle accidents are a random sample of fatal bicycle accidents. In addition, all expected counts in each category are all equal to $715\left(\frac{1}{8}\right) = 89.375$, which is greater than 5, so the sample size is large enough. The two conditions necessary to use the chi-square goodness-of-fit test have been satisfied.

Calculations (C):

Test statistic:

$$X^2 = \sum_{\text{all categories}} \frac{(\text{observed count} - \text{expected count})^2}{\text{expected count}}$$

$$= \frac{(38-89.375)^2}{89.375} + \frac{(29-89.375)^2}{89.375} + \cdots + \frac{(166-89.375)^2}{89.375} + \frac{(113-89.375)^2}{89.375}$$

$$= 29.5316 + 40.7848 + \cdots + 65.6939 + 6.2449$$

$$= 166.958$$

Degrees of freedom: $\text{df} = k - 1 = 8 - 1 = 7$

P-value: The P-value is the area under the chi-square curve with 7 degrees of freedom and to the right of 166.958. Therefore, the P-value is equal to $P(X^2 \geq 166.958) \approx 0$.

Communicate Results (C):

Because the P-value of approximately 0 is less than the chosen significance level of $\alpha = 0.05$, we reject H_0. There is convincing evidence that fatal accidents are not equally likely to occur in each of the eight time periods.

15.43: (a) Using the five-step process (HMC^3):

Hypotheses (H):

We want to determine if there is convincing evidence that fatal bicycle accidents are not equally likely to occur in each of the 12 months.

Define the following population characteristics of interest:

p_1 = proportion of fatal bicycle accidents in January

p_2 = proportion of fatal bicycle accidents in February

p_3 = proportion of fatal bicycle accidents in March

p_4 = proportion of fatal bicycle accidents in April

p_5 = proportion of fatal bicycle accidents in May

p_6 = proportion of fatal bicycle accidents in June

p_7 = proportion of fatal bicycle accidents in July

p_8 = proportion of fatal bicycle accidents in August

p_9 = proportion of fatal bicycle accidents in September

p_{10} = proportion of fatal bicycle accidents in October

p_{11} = proportion of fatal bicycle accidents in November

p_{12} = proportion of fatal bicycle accidents in December

State the appropriate hypotheses:

$$H_0 : p_1 = p_2 = p_3 = p_4 = p_5 = p_6 = p_7 = p_8 = p_9 = p_{10} = p_{11} = p_{12} = \frac{1}{12}$$

$$H_a : \text{At least one of the population proportions is not } \frac{1}{12}$$

Method (M):

Because the answers to the four key questions are (Q) hypothesis testing, (S) sample data, (T) one categorical variable with more than two categories, and (N) one sample, a chi-square goodness-of-fit test will be considered.

The test statistic for this test is

$$X^2 = \sum_{\text{all categories}} \frac{(\text{observed count} - \text{expected count})^2}{\text{expected count}}$$

When the null hypothesis is true, this statistic has approximately a chi-square distribution with $k - 1$ degrees of freedom.

A significance level of $\alpha = 0.01$ will be used for this test.

Check (C):

We are told (in exercise 15.41) to assume these bicycle accidents are a random sample of fatal bicycle accidents. In addition, all expected counts in each category are all equal to $719\left(\frac{1}{12}\right)=59.9167$, which is greater than 5, so the sample size is large enough. The two conditions necessary to use the chi-square goodness-of-fit test have been satisfied.

Calculations (C):

Test statistic:

$$X^2 = \sum_{\text{all categories}} \frac{(\text{observed count} - \text{expected count})^2}{\text{expected count}}$$
$$= \frac{(38-59.9167)^2}{59.9167} + \frac{(32-59.9167)^2}{59.9167} + \cdots + \frac{(42-59.9167)^2}{59.9167} + \frac{(40-59.9167)^2}{59.9167}$$
$$= 8.0168 + 13.0071 + \cdots + 5.3576 + 6.6204$$
$$= 82.1627$$

Degrees of freedom: $\text{df} = k-1 = 12-1 = 11$

P-value: The *P*-value is the area under the chi-square curve with 11 degrees of freedom and to the right of 82.1627. Therefore, the *P*-value is equal to $P(X^2 \geq 82.1627) \approx 0$.

Communicate Results (C):

Because the *P*-value of approximately 0 is less than the chosen significance level of $\alpha = 0.01$, we reject H_0. There is convincing evidence that fatal accidents are not equally likely to occur in each of the months.

(b)

$$H_0: p_1 = \frac{31}{366}, p_2 = \frac{29}{366}, p_3 = \frac{31}{366}, p_4 = \frac{30}{366}, p_5 = \frac{31}{366}, p_6 = \frac{30}{366},$$
$$p_7 = \frac{31}{366}, p_8 = \frac{31}{366}, p_9 = \frac{30}{366}, p_{10} = \frac{31}{366}, p_{11} = \frac{30}{366}, p_{12} = \frac{31}{366}$$

H_a: At least one of the population proportions is not correct as written

(c) Using the five-step process (HMC^3):

Hypotheses (H):

We want to determine if there is convincing evidence that fatal bicycle accidents are not equally likely to occur in each of the 12 months in proportion to the lengths of the months.

Define the following population characteristics of interest:

p_1 = proportion of fatal bicycle accidents in January

p_2 = proportion of fatal bicycle accidents in February

p_3 = proportion of fatal bicycle accidents in March

p_4 = proportion of fatal bicycle accidents in April

p_5 = proportion of fatal bicycle accidents in May

p_6 = proportion of fatal bicycle accidents in June

p_7 = proportion of fatal bicycle accidents in July

p_8 = proportion of fatal bicycle accidents in August

p_9 = proportion of fatal bicycle accidents in September

p_{10} = proportion of fatal bicycle accidents in October

p_{11} = proportion of fatal bicycle accidents in November

p_{12} = proportion of fatal bicycle accidents in December

State the appropriate hypotheses:

$$H_0: p_1 = \frac{31}{366}, p_2 = \frac{29}{366}, p_3 = \frac{31}{366}, p_4 = \frac{30}{366}, p_5 = \frac{31}{366}, p_6 = \frac{30}{366},$$
$$p_7 = \frac{31}{366}, p_8 = \frac{31}{366}, p_9 = \frac{30}{366}, p_{10} = \frac{31}{366}, p_{11} = \frac{30}{366}, p_{12} = \frac{31}{366}$$

H_a: At least one of the population proportions is not correct as written

Method (M):

Because the answers to the four key questions are (Q) hypothesis testing, (S) sample data, (T) one categorical variable with more than two categories, and (N) one sample, a chi-square goodness-of-fit test will be considered.

The test statistic for this test is

$$X^2 = \sum_{\text{all categories}} \frac{(\text{observed count} - \text{expected count})^2}{\text{expected count}}$$

When the null hypothesis is true, this statistic has approximately a chi-square distribution with $k - 1$ degrees of freedom.

A significance level of $\alpha = 0.05$ will be used for this test.

Check (C):

We are told (in exercise 15.41) to assume these bicycle accidents are a random sample of fatal bicycle accidents. In addition, the table below shows the expected counts in each category, which are all greater than 5, so the sample size is large enough.

Month	Expected Counts	Month	Expected Counts
January	$719\left(\frac{31}{366}\right) = 60.8989$	July	$719\left(\frac{31}{366}\right) = 60.8989$
February	$719\left(\frac{29}{366}\right) = 56.9699$	August	$719\left(\frac{31}{366}\right) = 60.8989$
March	$719\left(\frac{31}{366}\right) = 60.8989$	September	$719\left(\frac{30}{366}\right) = 58.9344$
April	$719\left(\frac{30}{366}\right) = 58.9344$	October	$719\left(\frac{31}{366}\right) = 60.8989$
May	$719\left(\frac{31}{366}\right) = 60.8989$	November	$719\left(\frac{30}{366}\right) = 58.9344$
June	$719\left(\frac{30}{366}\right) = 58.9344$	December	$719\left(\frac{31}{366}\right) = 60.8989$

The two conditions necessary to use the chi-square goodness-of-fit test have been satisfied.

Calculations (C):

Test statistic:

$$X^2 = \sum_{\text{all categories}} \frac{(\text{observed count} - \text{expected count})^2}{\text{expected count}}$$

$$= \frac{(38-60.8989)^2}{60.8989} + \frac{(32-56.9699)^2}{56.9699} + \cdots + \frac{(42-58.9344)^2}{58.9344} + \frac{(40-60.8989)^2}{60.8989}$$

$$= 8.6103 + 10.9443 + \cdots + 4.8660 + 7.1720$$

$$= 78.5105$$

Degrees of freedom: $\text{df} = k-1 = 12-1 = 11$

P-value: The *P*-value is the area under the chi-square curve with 11 degrees of freedom and to the right of 78.5105. Therefore, the *P*-value is equal to $P(X^2 \geq 78.5105) \approx 0$.

Communicate Results (C):

Because the *P*-value of approximately 0 is less than the chosen significance level of $\alpha = 0.05$, we reject H_0. There is convincing evidence that fatal bicycle accidents do not occur in the twelve months in proportion to the lengths of the months.

15.45: Using the five-step process (HMC^3):

Hypotheses (H):

We want to determine if there is convincing evidence that the spirituality category proportions are not the same for natural and social scientists.

The hypotheses of interest are:

H_0: The proportions falling into the spirituality categories are the same for natural and social scientists

H_a: The proportions falling into the spirituality categories are not the same for natural and social scientists

Method (M):

The purpose of this study is to compare two populations (natural scientists and social scientists) on the basis of one categorical variable (degree of spirituality), so a chi-square test of homogeneity will be considered.

The test statistic for this test is

$$X^2 = \sum_{\text{all categories}} \frac{(\text{observed count} - \text{expected count})^2}{\text{expected count}}$$

When the null hypothesis is true, this statistic has approximately a chi-square distribution with (number of rows – 1)(number of columns – 1) degrees of freedom.

A significance level of $\alpha = 0.01$ will be used for this test.

Check (C):

We are told that random samples of both natural and social scientists were selected. The table below shows observed counts and, in parentheses, expected counts. Note that all expected counts are at least 5.

	Degree of Spirituality			
	Very	Moderate	Slightly	Not at All
Natural Scientists	56 (50.59)	162 (173.92)	198 (199.21)	211 (203.28)
Social Scientists	56 (61.41)	223 (211.08)	243 (241.79)	239 (246.72)

Because the samples were independent random samples and all expected counts are at least 5, the chi-square test of homogeneity is appropriate.

Calculations (C):

Test statistic:

$$X^2 = \sum_{\text{all categories}} \frac{(\text{observed count} - \text{expected count})^2}{\text{expected count}}$$

$$= \frac{(56-50.59)^2}{50.59} + \frac{(162-173.92)^2}{173.92} + \cdots + \frac{(243-241.79)^2}{241.79} + \frac{(239-246.72)^2}{246.72}$$

$$= 0.578 + 0.816 + \cdots + 0.006 + 0.242$$

$$= 3.091$$

Degrees of freedom:

$\text{df} = (\text{number of rows} - 1)(\text{number of columns} - 1) = (1)(3) = 3$

P-value: The *P*-value is the area under the chi-square curve with 3 degrees of freedom and to the right of 3.091. Therefore, the *P*-value is equal to $P(X^2 \geq 3.091) = 0.378$.

Communicate Results (C):

Because the *P*-value of 0.378 is greater than the specified significance level of $\alpha = 0.01$, we fail to reject H_0. There is not convincing evidence that the spirituality category proportions are not all the same for natural scientists and social scientists.

15.47: Using the five-step process (HMC^3):

Hypotheses (H):

We want to determine if there is convincing evidence that there is an association between whether or not children are overweight 1 year after and the number of sweet drinks consumed.

The hypotheses of interest are:

H_0: There is no association between whether or not a child is overweight and the number of sweet drinks consumed

H_a: There is an association between whether or not a child is overweight and the number of sweet drinks consumed

Method (M):

The purpose of this study is to compare two categorical variables (overweight and number of sweet drinks consumed) from one population (children who were underweight or normal weight at age 2), so a chi-square test of association will be considered.

The test statistic for this test is

$$X^2 = \sum_{\text{all categories}} \frac{(\text{observed count} - \text{expected count})^2}{\text{expected count}}$$

When the null hypothesis is true, this statistic has approximately a chi-square distribution with (number of rows – 1)(number of columns – 1) degrees of freedom.

A significance level of $\alpha = 0.05$ will be used for this test.

Check (C):

We are told to assume that the sample of children in this study is representative of 2- to 3-year-old children. The table below shows observed counts and, in parentheses, expected counts. Note that all expected counts are at least 5.

		Overweight?	
		Yes	No
Number of Sweet Drinks Consumed per Day	0	22 (28.92)	930 (923.08)
	1	73 (65.22)	2,074 (2,081.78)
	2	56 (52.77)	1,681 (1,684.23)
	3 or More	102 (106.09)	3,390 (3,385.91)

Because we are assuming that this is a representative sample and all expected counts are at least 5, the chi-square test of association is appropriate.

Calculations (C):

Test statistic:

$$X^2 = \sum_{\text{all categories}} \frac{(\text{observed count} - \text{expected count})^2}{\text{expected count}}$$

$$= \frac{(22-28.92)^2}{28.92} + \frac{(930-923.08)^2}{923.08} + \cdots + \frac{(102-106.09)^2}{106.09} + \frac{(3{,}390-3{,}385.91)^2}{3{,}385.91}$$

$$= 1.656 + 0.052 + \cdots + 0.157 + 0.005$$

$$= 3.030$$

Degrees of freedom:
$\text{df} = (\text{number of rows} - 1)(\text{number of columns} - 1) = (3)(1) = 3$

P-value: The *P*-value is the area under the chi-square curve with 3 degrees of freedom and to the right of 3.030. Therefore, the *P*-value is equal to $P(X^2 \geq 3.030) = 0.387$.

Communicate Results (C):

Because the *P*-value of 0.387 is greater than the specified significance level of $\alpha = 0.05$, we fail to reject H_0. There is not convincing evidence that there is an association between whether or not a child is overweight and the number of sweet drinks consumed.

15.49: (a) H_0: The proportions falling in the water consumption categories are the same for males and females versus H_a: The proportions falling in the water consumption categories are not the same for males and females. Degrees of freedom are calculated using the formula (number of rows – 1)(number of columns – 1). Since there are two rows corresponding to gender (male, female), and five columns corresponding to number of servings of water consumed (none, one, two to three, four to five, six or more), there are (2 – 1)(5 – 1) = (1)(4) = 4 degrees of freedom.

(b) The *P*-value for the test was 0.086, which is greater than the new significance level of 0.05. So, for a significance level of 0.05, we do not have convincing evidence of a difference between males and females with regard to water consumption.

15.51: (a)

	Napped	Did Not Nap	Row Total
Men	283	461	**744**
Women	231	513	**744**

(b) Using the five-step process (HMC^3):

Hypotheses (H):

We want to determine if there is convincing evidence that there is an association between gender and napping.

The hypotheses of interest are:

H_0: There is no association between gender and napping

H_a: There is an association between gender and napping

Method (M):

The purpose of this study is to compare two categorical variables (gender and napping) from one population (adult Americans), so a chi-square test of association will be considered.

The test statistic for this test is

$$X^2 = \sum_{\text{all categories}} \frac{(\text{observed count} - \text{expected count})^2}{\text{expected count}}$$

When the null hypothesis is true, this statistic has approximately a chi-square distribution with (number of rows – 1)(number of columns – 1) degrees of freedom.

A significance level of $\alpha = 0.05$ will be used for this test.

Check (C):

We are told to assume that the sample is representative of adult American. The table below shows observed counts and, in parentheses, expected counts. Note that all expected counts are at least 5.

	Napped	Did Not Nap
Men	283 (257)	461 (487)
Women	231 (257)	513 (487)

Because the sample is representative and all expected counts are at least 5, the chi-square test of association is appropriate.

Calculations (C):

Test statistic:

$$X^2 = \sum_{\text{all categories}} \frac{(\text{observed count} - \text{expected count})^2}{\text{expected count}}$$

$$= \frac{(283-257)^2}{257} + \frac{(461-487)^2}{487} + \frac{(231-257)^2}{257} + \frac{(513-487)^2}{487}$$

$$= 2.630 + 1.388 + 2.630 + 1.388$$

$$= 8.037$$

Degrees of freedom:
$\text{df} = (\text{number of rows} - 1)(\text{number of columns} - 1) = (1)(1) = 1$

P-value: The P-value is the area under the chi-square curve with 1 degree of freedom and to the right of 8.037. Therefore, the P-value is equal to $P(X^2 \geq 8.037) = 0.005$.

Communicate Results (C):

Because the P-value of 0.005 is less than the selected significance level of $\alpha = 0.05$, we reject H_0. There is convincing evidence that there is an association between gender and napping.

(c) Yes. First, recall that the sample is representative of all adult Americans, so the conclusions drawn from this study can be generalized to the population of all adult Americans. Second, the results of this study provide evidence of an association between gender and napping in the population. The number who napped was greater than expected for men and less than expected for women.

CPSIA information can be obtained
at www.ICGtesting.com
Printed in the USA
FFOW021734230113
758FF

9 781285 08983